SYMMETRY IN CHEMICAL BONDING AND STRUCTURE

MERRILL CHEMISTRY SERIES

THEODORE L. BROWN, EDITOR

SYMMETRY IN CHEMICAL BONDING AND STRUCTURE

WILLIAM E. HATFIELD
University of North Carolina at Chapel Hill

WILLIAM E. PARKER
Gettysburg College

Published by
Charles E. Merrill Publishing Co.
A Bell & Howell Company
Columbus, Ohio 43216

International Standard Book Number: 0-675-08931-x
Library of Congress Card Catalog Number: 73-76075
1 2 3 4 5 6 7 8—78 77 76 75 74

Printed in the United States of America

PREFACE

Recently there has been a great deal of use of symmetry and group theory in chemistry. The applications range from the determination of structures by application of spectroscopic selection rules to the description of reaction mechanisms. In fact, the use of symmetry notation is so commonplace that it is nearly impossible to read the current chemical literature without some knowledge of the notation. For example, recent advances in instrumentation and computational facilities permit the determination of molecular and crystal structures in a relatively short time, and the number of structure descriptions in the journals is increasing at a rapid rate. Yet many chemists are not familiar with the notation adopted by crystallographers. With these ideas in mind we have written this book in which we have described some of the notation and developed some of the concepts of symmetry which are useful in problems with chemical bonding and structure. We have also introduced some of these latter topics. This book is for advanced undergraduates and beginning graduate students and is intended only as an introduction to the material. Consequently, we have not attempted an exhaustive development even of the topics that we have covered, nor have we presented much experimental data and results. We have provided a number of exercises which serve to test the understanding of the material and which also extend the coverage of the topics. The discussions in this book should stimulate additional reading and study in appropriate monographs, from which we have drawn much of the material, and for which we have provided references at the end of appropriate chapters. To the authors of these books we owe a debt of gratitude.

We wish to thank Professors Carol Gatz, Marcetta Y. Darensbourg, and Donald J. Darensbourg, who read the manuscript, for their helpful comments, and Mr. David C. Bull of W. A. Benjamin, Inc. for permission to use material from *Problems in Structural Inorganic Chemistry* by W. E. Hatfield and R. A. Palmer.

William E. Hatfield
William E. Parker

CONTENTS

1

SYMMETRY ELEMENTS AND OPERATIONS

Every person is constantly exposed to symmetrical and unsymmetrical objects, though often unaware of it. In the preparation of a report or paper, the writer is careful to center the title and the text on the pages of the final copy. A chemist looks at the structure of molecules like benzene or nitrobenzene and notes that there is only one type of hydrogen atom in benzene and three types of hydrogen atoms in nitrobenzene. In this chapter we will introduce some basic ideas about symmetry elements and symmetry operations which will help systematize the principles of symmetry.

1-1 DEFINITIONS

Although closely related, the terms symmetry element and symmetry operation represent two different things, and their definitions should be kept firmly in mind throughout the use of this book.

Symmetry Operation

A *symmetry operation* is a movement of an object such that, after the movement has been performed, every point of the object is coincident with an equivalent point in the original orientation.

Symmetry Element

A *symmetry element* is a line, point, or plane about which a symmetry operation may be performed.

The close relationship between symmetry element and symmetry operation is now apparent; the motion of the symmetry operation is always going to be guided by the appropriate symmetry element.

1-2 DESCRIPTIONS OF ELEMENTS AND OPERATIONS

Identity

The identity element, designated by E, is an element that all molecules or objects must possess. The operation that can be performed with respect to E is really no operation at all, because this operation not only produces an equivalent orientation, but an identical one.

Proper Axis of Rotation

A proper axis of rotation, designated by the symbol C, is an imaginary line contained within the object such that a rotation by a specified number of degrees will produce an equivalent orientation of the object. In order to specify the number of degrees of rotation, a subscript n is incorporated in the symbol. The symbol designating a proper rotation is C_n, where n equals 360°/(number of degrees of rotation) and is called the *order* of the rotation axis. For example, if rotation of 180° about the axis C produces an equivalent orientation of the object, n must equal 2, and we have a C_2 axis of rotation. If rotation of 120° produces an equivalent orientation, n must equal 3, and we have a C_3 axis of rotation. A C_1 axis of rotation is equivalent to the identity element, because a rotation of 360° produces not only an equivalent orientation but an identical one ($C_1 = \mathrm{E}$).

The same basic symbol is used to denote both the symmetry element and the symmetry operation, although successive rotations are frequently denoted by a post superscript. Examples which illustrate the rotation are given below. Although the same symbol is used for both operations and elements, little confusion need arise since the intent of the symbol is usually clear from the context.

In Figure 1.1 the trigonal planar molecule BF_3 contains three identical fluorine atoms bonded to a central boron atom, and the C_3 rotation axis is perpendicular to the plane of the molecule and passes through the boron atom. The fluorine atoms are numbered so that we can identify them after each operation is performed. Note that there are two possible rotations of 120° before returning to the original orientation.

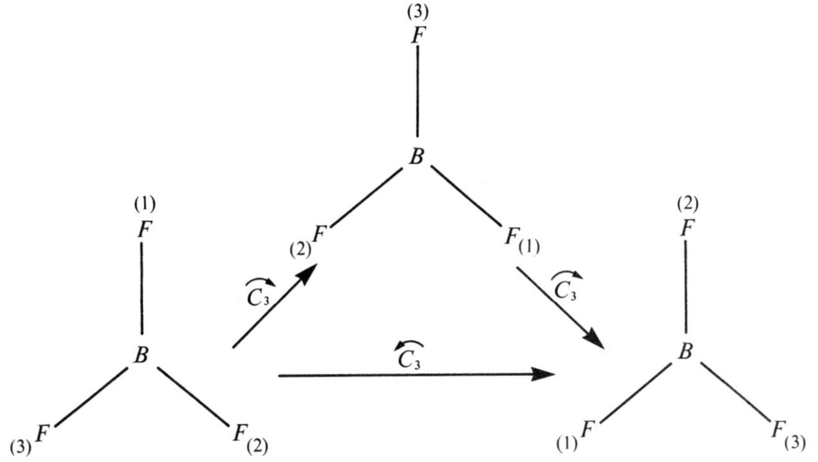

Figure 1.1
C_3 Rotations of BF_3

A molecule may contain only one axis of rotation, or it may contain many axes of the same or of different orders. If again we consider the BF_3 molecule, we can see in Figure 1.2 that there are three C_2 axes, which are in the plane of the molecule and are perpendicular to the C_3 axis. When a molecule possesses rotation axes of different orders, the axis that has the highest order is called the principal axis of rotation. If the molecule contains more than one axis of the highest order, the principal one is taken to be colinear with a unique molecular axis.

We can perform the rotation operation by rotating either in a clockwise direction or in a counterclockwise direction. When the rotation axis is of first or second order (C_1 or C_2), the direction of rotation makes no difference, because we arrive at the same equivalent orientation of the molecule in either case. However, for rotation axes of order greater than 2, such as the C_3 axis in BF_3, the resultant orientations of the molecule depend upon the direction of rotation. In Figure 1.3 we can see that a rotation of 120° about the C_3 axis in a clockwise direction does not produce the same result as a rotation of 120° in a counterclockwise direction. However, rotating 240° in one direction or rotating 120° in the opposite direction produces identical orientations. Three different operations with respect to the C_3 axis of rotation exist: if we rotate 120° in a clockwise direction we obtain one equivalent orientation; if we rotate 120° in a counterclockwise direction, or 240° in a clockwise direction, we obtain a second equivalent orientation; and if we rotate 360° in either direction we have the original

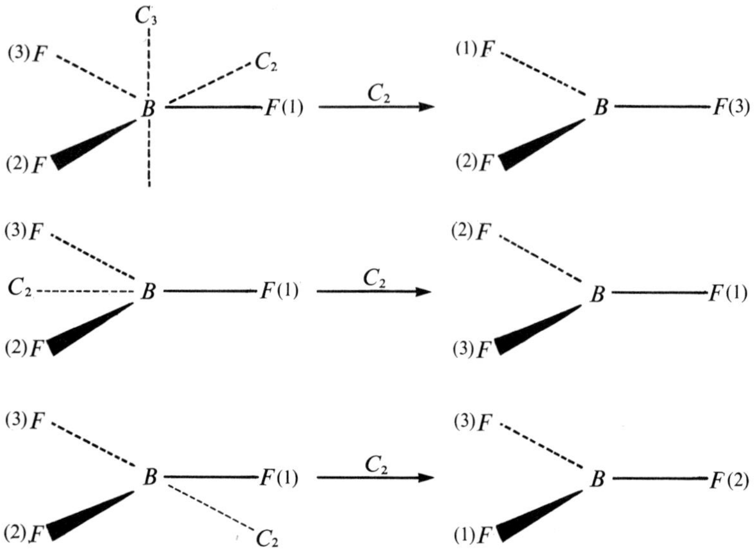

Figure 1.2
C_2 Rotations of BF_3

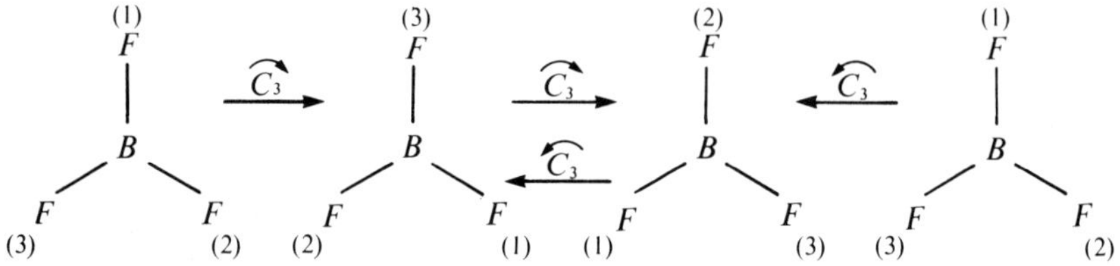

Figure 1.3
C_3 Rotations of BF_3

orientation of the molecule. Symbolically, $\overrightarrow{C}_3$, $\overleftarrow{C}_3 = \overrightarrow{C}_3^2$, and $\overleftarrow{C}_3^3 = \overrightarrow{C}_3^3 = E$ are the three operations that we can perform about the C_3 axis of rotation. The identity element and the identity operation have already been discussed, and because $C_3^3 = E$, there are only two unique operations that can be performed about a C_3 axis. These are $\overrightarrow{C}_3$ and $\overleftarrow{C}_3$. A word about unique operations: Occasionally, the same orientation of an object will result from the application of two distinctly different operations. We shall see later that it is important to recognize this duplication and to account for it.

To further illustrate rotation operations we will consider the square planar anion $PtCl_4^{2-}$. In this ion there is a C_4 axis of rotation perpendicular to the plane of the ion and passing through the platinum atom. As shown in Figure 1.4 upon application of a single C_4 operation we get two different equivalent orientations depending on the direction of rotation. If we rotate 270° in one direction or 90° in the opposite direction, we get identical results. Upon further inspection we find five C_2 axes of rotation, as illustrated in Figure 1.5. Four of these axes lie in the plane of the ion and

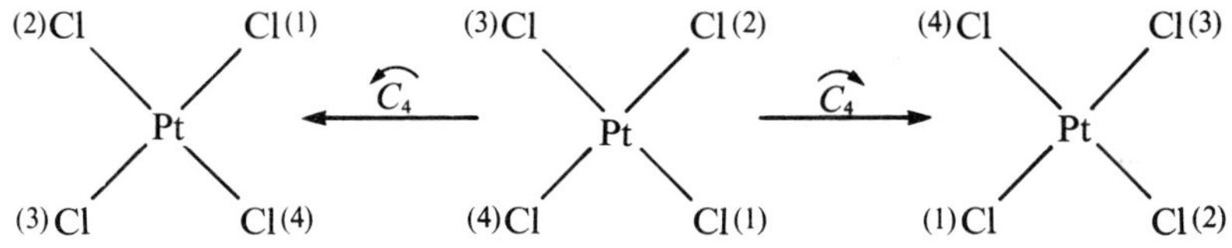

Figure 1.4
C_4 Rotations of $PtCl_4^{2-}$

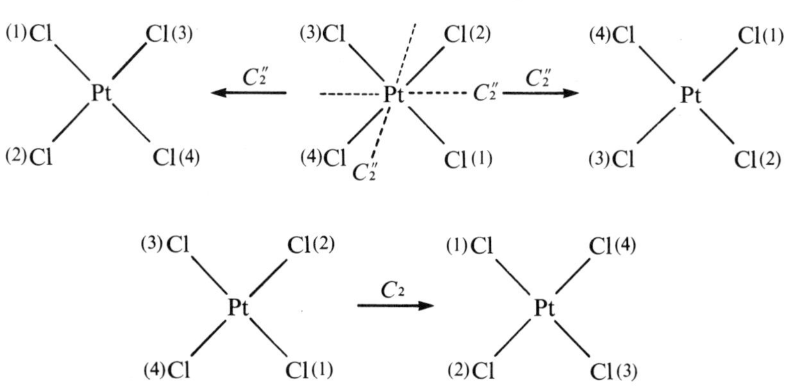

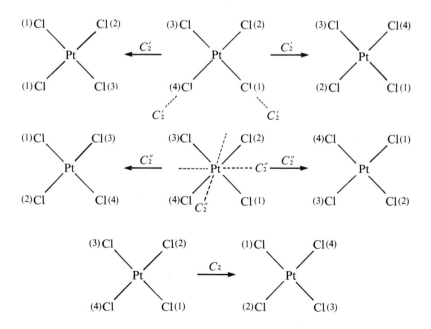

Figure 1.5
C_2 Rotations of $PtCl_4^{2-}$

are designated C'_2 or C''_2, depending on their location; the fifth C_2 axis is colinear with the C_4 principal axis. For the $PtCl_4^{2-}$ ion a 90° rotation repeated four successive times in the same direction reproduces the original orientation, and therefore $C_4^4 = E$. A 90° rotation repeated two successive times in the same direction about the C_4 axis produces the same result as a 180° rotation about the C_2 axis, which is colinear with the C_4 axis, and therefore $C_4^2 = C_2$. The unique operations that are present for the $PtCl_4^{2-}$ ion, involving either the identity element or rotation axes, are $E = C_4^4$, $\overrightarrow{C}_4$, $\overleftarrow{C}_4$, $C_2 = C_4^2$, two C'_2, and two C''_2.

We have discussed first-, second-, third-, and fourth-order axes up to this point, but n may have any integral value up to and including infinity. A linear molecule, such as H_2, which is shown in Figure 1.6, has a rotation

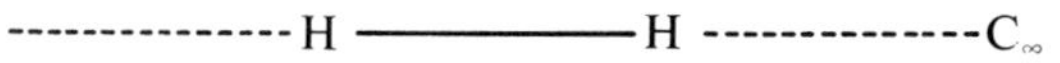

Figure 1.6
C_∞ Rotation of H_2

axis colinear with the molecular axis. Because a rotation by an infinitesimal number of degrees about this axis will produce an identical orientation of the molecule, n must equal infinity, and the principal axis of rotation is designated C_∞. Figure 1.7 shows various examples of proper axes of rotation and illustrates one symmetry operation about each axis.

Plane of Symmetry or Mirror Plane

A plane of symmetry, designated by σ, is a plane that passes through the molecule and bisects that molecule into halves that are mirror images of each other. When the operation of reflection is performed with respect to a plane of symmetry, the result is an orientation equivalent to the original orientation of the molecule.

We must learn to identify three different types of symmetry planes. The first of these is a horizontal plane of symmetry, designated by σ_h, which lies perpendicular to the principal axis of rotation of the molecule. Figure 1.8 shows that in the BF_3 molecule there is a plane of symmetry (σ_h) which includes all of the atoms of the molecule and is perpendicular to the principal (C_3) axis. The $C_5H_5^-$ ion has a principal axis of order 5 (C_5), and perpendicular to this axis lies a plane (σ_h) that includes all of the atoms of the anion. The octahedral MF_6^{3-} ion has three axes of order 4 (C_4), and any one of them can be designated as the principal axis of rotation. There

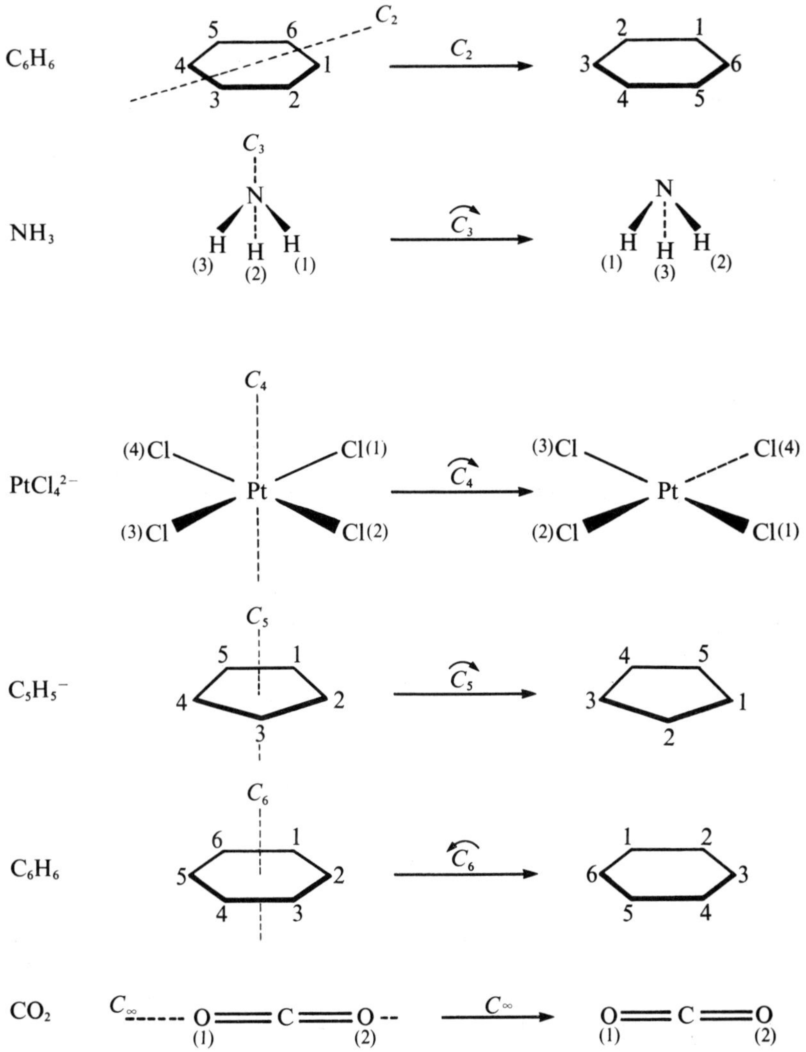

Figure 1.7
Proper Axes of Rotation

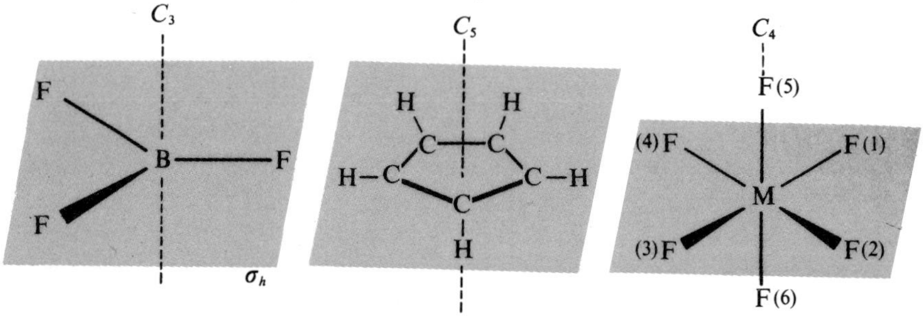

Figure 1.8
Horizontal Planes of Symmetry

are three planes of symmetry (σ_h), each of which is perpendicular to the three C_4 axes; one of these is illustrated in Figure 1.8. The MF_6^{3-} ion is the only example in Figure 1.8 where the reflection operation actually changes the positions of any atoms; the other two examples are planar. In the MF_6^{3-} ion, the fluorine ligands 5 and 6 are exchanged, while 1 through 4 remain where they were, since they lie in the σ_h plane.

A vertical plane of symmetry, designated by σ_v, must include the principal axis of rotation. If no symmetry axis of rotation exists in the molecule, then any existing plane is defined as a σ_v plane of symmetry. In Figure 1.9 the $CHCl_2Br$ molecule has no proper axis of rotation but does have the plane of symmetry illustrated. Reflection across this plane interchanges the two chlorine atoms. The BF_3 molecule, as already pointed out, has a

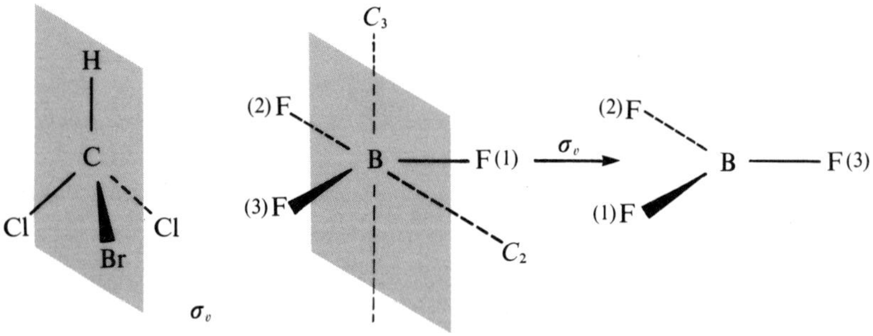

Figure 1.9
Vertical Planes of Symmetry

C_3 principal axis of rotation, three C_2 axes of rotation that are perpendicular to the C_3 axis, and a σ_h plane perpendicular to the C_3 axis. There are three σ_v planes in BF_3; one of these is illustrated in Figure 1.9.

A σ_d, or dihedral plane of symmetry is very similar to a σ_v plane of symmetry, because the principal rotation axis of the molecule still must lie within the plane; but these also bisect the angles between two C_2 axes that are perpendicular to the major axis.

Figure 1.10 shows a few of the symmetry elements of the $PtCl_4^{2-}$ anion. Two of the vertical planes, which are arbitrarily called σ_v planes of symmetry, include the two C_2 axes that pass through the chlorine atoms, while the other two vertical planes, which are σ_d planes of symmetry, bisect the angles between the two C_2 axes that are illustrated.

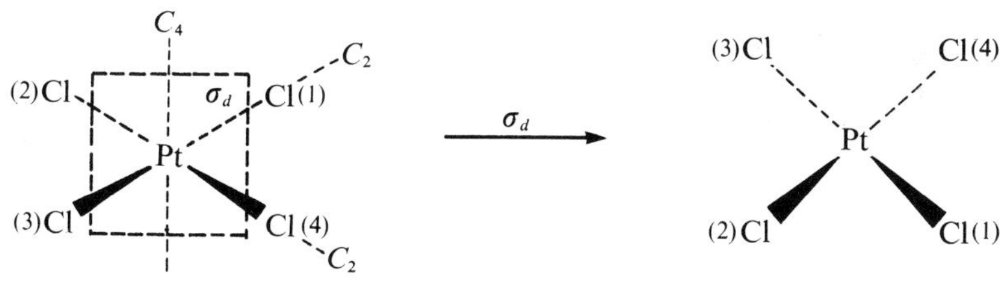

Figure 1.10
Symmetry Elements of $PtCl_4^{2-}$

Any linear molecule, such as CO_2, has an infinite number of planes of symmetry which include the principal axis of rotation (C_∞); therefore, these planes are all vertical planes of symmetry (σ_v).

Center of Symmetry or Inversion Center

A center of symmetry, designated by *i*, exists in a molecule if inversion of each of the atoms through this inversion center results in an equivalent orientation of the molecule. In other words, except for an atom coincident with the center, atoms must occur in pairs and be equidistant and in opposite directions from the center of the molecule. The examples shown in Figure 1.11 illustrate the operation associated with the inversion element.

A tetrahedral molecule, such as CCl_4, does not have a center of symmetry, although the carbon atom lies at the center of the molecule. No center of symmetry is present, because inversion of the four chlorine atoms through the carbon atom does not result in an equivalent orientation of the molecule. The molecules CO_2 and C_6H_6 and the anion $PtCl_4^{2-}$ have

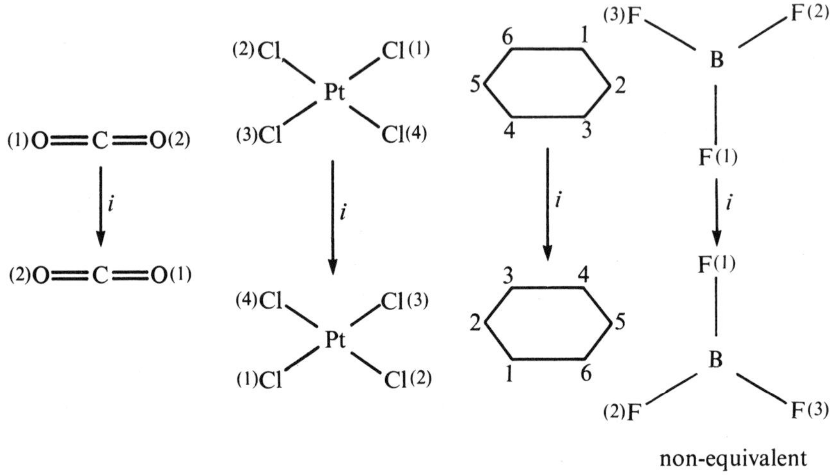

Figure 1.11
Centers of Symmetry

centers of inversion (i) as illustrated in Figure 1.11, but BF_3 has no such center.

Improper Axis of Rotation

An improper axis of rotation, designated by S, is similar to the proper axis of rotation (C). The difference is that after the rotation operation, we must perform a reflection across a plane which is perpendicular to the rotation axis. This dual operation is called an improper rotation. This represents the only case where a combination of operations (C followed by σ) leads to a new operation (S). The subscript n, representing the order of the axis, is used in the same way for both C and S axes of rotation. The rotation axis and the reflection plane need not be true symmetry elements of the molecule. The tetrahedral CCl_4 molecule, illustrated in Figure 1.12, possesses an S_4 symmetry axis, but not a C_4 symmetry axis. In Figures 1.12 and 1.13 operations which are not in themselves true symmetry operations have been put in brackets to distinguish them from true symmetry operations.

Just as a C_1 axis of rotation is trivial ($C_1 = E$), an S_1 axis is equivalent to an operation that we have already discussed. A rotation of 360° followed by a reflection through a plane perpendicular to the axis of rotation yields the same result as the reflection performed alone. In other words, every

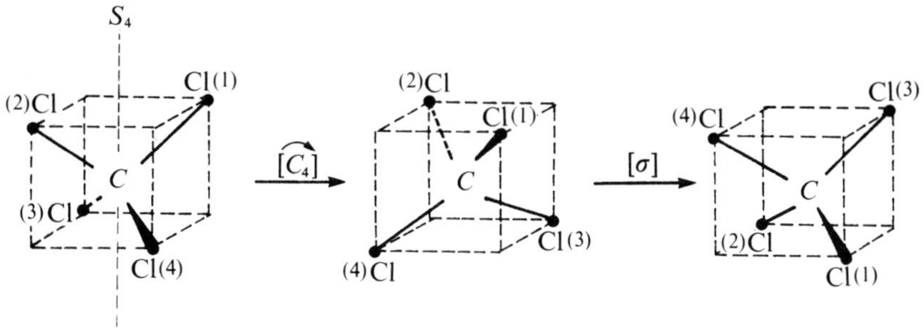

Figure 1.12
S_4 Symmetry Operation

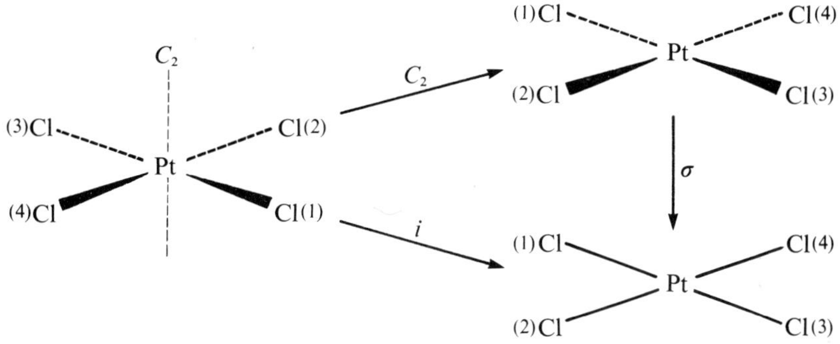

Figure 1.13
S_2 Rotations

molecule which possesses a plane of symmetry also must have an S_1 axis perpendicular to that plane.

An S_2 rotation is equivalent to the inversion operation ($S_2 = i$), as shown in Figure 1.13. Any molecule containing a center of symmetry also possesses an infinite number of lines which pass through the inversion center and represent S_2 axes of rotation.

When the improper axis of rotation has an order greater than 2, clockwise and counterclockwise rotations represent different operations with respect to the same element. For example, we may perform four successive operations with respect to an S_4 axis: $\overrightarrow{S}_4$, $\overrightarrow{S}_4^2 = C_2$, $\overrightarrow{S}_4^3 = \overleftarrow{S}_4$, and $\overrightarrow{S}_4^4 = E$;

Table 1-1
Symmetry Elements and Operations

Symbol	*Element*	*Operation*
E	identity	
C	proper axis of rotation	rotation
σ	plane	reflection
i	center of symmetry	inversion
S	improper axis of rotation	rotation followed by reflection

however, only two of these represent operations that are different from the ones we have already discussed. When an operation is performed an even number of times with respect to an improper axis of rotation, the reflection component of the operations has no effect on the orientation of the molecule ($\sigma^2 = E$); because the operation C_4^2 equals the C_2 operation, an S_4^2 operation must be equivalent to a C_2 operation. Just as in the case of the proper axis of rotation, more than one improper axis of rotation of the same order may exist in a molecule. The symmetry elements and operations with which we should be familiar are collected in Table 1.1.

1-3 IDENTIFICATION OF SYMMETRY ELEMENTS

Now that we have defined the basic symmetry elements and described the symmetry operations that can be performed with respect to these symmetry elements, we are ready to identify all the symmetry elements present in specific molecules.

The NH_3 molecule, shown in Figure 1.14, has a trigonal pyramidal

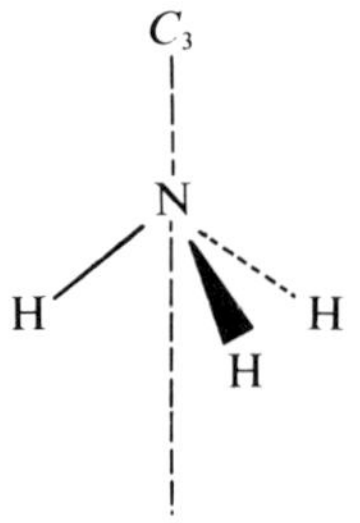

Figure 1.14
The NH_3 *Molecule*

structure and, of course, must possess an identity element E. There is a proper axis of rotation with the order 3, i.e., a C_3 axis. Because the molecule has a pyramidal structure, there are no C_2 axes of rotation which are perpendicular to the C_3 axis. No σ_h plane of symmetry is present, but there are three σ_v planes of symmetry which include the C_3 principal axis and one of the three hydrogen atoms. The NH_3 molecule does not possess a center of inversion or any improper axes of rotation. Therefore, the only symmetry elements present in the NH_3 molecule are E, C_3, and three σ_v.

The $PtCl_4^{2-}$ anion was discussed earlier in this chapter. This square planar ion possesses an identity element, and, as illustrated in Figure 1.15, the rotation axis of highest order is a C_4 axis perpendicular to the plane of the molecule. As pointed out when we were discussing proper axes of rotation, there are five C_2 axes in this structure. One of these axes is colinear with the C_4 axis of rotation, and the other four lie perpendicular to the principal axis. Because $PtCl_4^{2-}$ is planar, a horizontal plane of symmetry (σ_h) must exist. In addition, two σ_v planes that each include two chlorine atoms, and two σ_d planes that bisect the angles between any two adjacent chlorine atoms are present. The ion $PtCl_4^{2-}$ also contains a center of symmetry (i), which is coincident with the platinum atom, and an S_4 axis of rotation which is colinear with the C_4 axis. The $PtCl_4^{2-}$ ion possesses the following symmetry elements: E, C_4, five C_2, σ_h, two σ_v, two σ_d, i, and S_4.

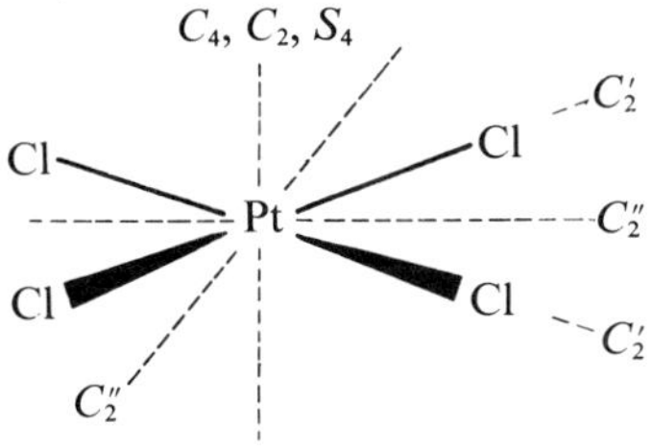

Figure 1.15
Rotation Axes of $PtCl_4^{2-}$

The BF_3 molecule, illustrated in Figure 1.16, has a trigonal planar structure. In addition to the identity element, a C_3 axis of rotation perpendicular to the plane of the molecule is present and is the principal axis of rotation. Three C_2 axes lie perpendicular to the principal axis, and an S_3 axis of rotation colinear with the C_3 axis is also present. A horizontal plane of symmetry (σ_h), which includes all the atoms of the molecule, and three vertical planes of symmetry (σ_v), each of which includes the C_3 principal axis and one of the perpendicular C_2 axes, are among the symmetry

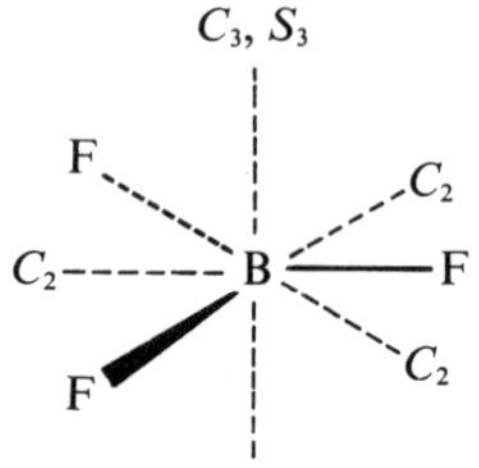

Figure 1.16
Rotation Axes of BF_3

elements of the BF_3 molecule. No inversion center is present in this molecule. The symmetry elements for the BF_3 molecule are E, C_3, three C_2, S_3, σ_h, and three σ_v.

The PF_5 molecule has the trigonal bipyramidal structure shown in Figure 1.17. This molecule is similar to the BF_3 molecule because the principal axis of rotation is a C_3 axis. In addition to the identity element, the PF_5 molecule contains three C_2 axes perpendicular to the C_3 principal axis, and an S_3 axis colinear with the C_3 axis. A horizontal plane of symmetry (σ_h) and three vertical planes of symmetry (σ_v) are also recognizable. The BF_3 and PF_5 molecules have exactly the same symmetry elements (E, C_3, three C_2, S_3, σ_h, and three σ_v).

The CO_2 molecule is linear and thus has a C_∞ axis of rotation as shown in Figure 1.18. There are an infinite number of C_2 axes, which pass through the carbon atom and are perpendicular to the C_∞ axis of rotation, and an

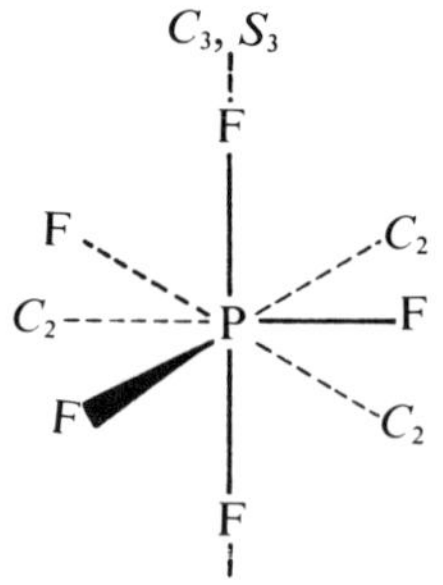

Figure 1.17
Rotation Axes of PF_5

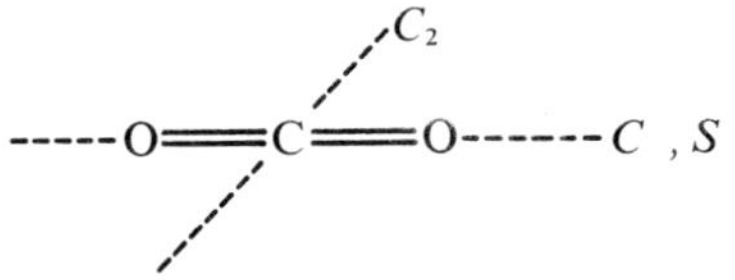

Figure 1.18
Rotation Axes of CO_2

S_∞ axis that is colinear with the C_∞ axis. In this particular molecule a horizontal plane of symmetry is equivalent to an improper axis of rotation ($\sigma_h = S_\infty$). An infinite number of σ_v planes, which include the C_∞ principal axis, and a center of inversion, which is coincident with the carbon atom, are also present. The symmetry elements contained in the CO_2 molecule are E, C_∞, an infinite number of C_2 axes, an infinite number of σ_v planes, and i.

The NCS^- anion is also linear, but since there are different atoms bonded to the central atom, we will probably find fewer symmetry elements than in CO_2. Along with the identity element, a C_∞ axis of rotation is present, and is illustrated in Figure 1.19. An infinite number of σ_v planes of symmetry are present, as in the case of CO_2, but no center of symmetry, no σ_h, no C_2 axes of rotation, and no S_∞ axis of rotation exist in the NCS^- ion. Therefore, the only symmetry elements present in this anion are E, C_∞, and an infinite number of σ_v planes.

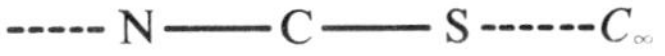

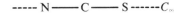

Figure 1.19
Rotation Axis of NCS^-

The $POCl_3$ molecule has a tetrahedral geometry as illustrated in Figure 1.20. The principal axis of rotation is a C_3 axis that is colinear with the P=O bond. No other axes of rotation and no center of inversion are present. No horizontal plane of symmetry exists, but three vertical planes, each of which includes the C_3 principal axis and one of the chlorine atoms, are symmetry elements of the $POCl_3$ molecule. The symmetry elements present in this molecule are the same as for the NH_3 molecule (E, C_3, and three σ_v).

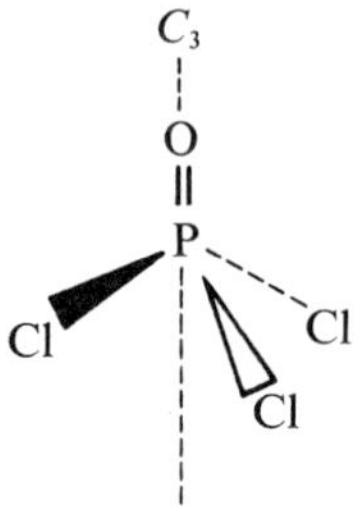

Figure 1.20
C_3 Axis of $POCl_3$

The molecules and ions discussed in this section represent a few examples that illustrate the procedure involved in identifying symmetry elements.

1-4 COMBINATIONS OF SYMMETRY OPERATIONS

Of all the symmetry elements that have been discussed, the only one that involves more than one symmetry operation is the improper axis of rotation ($S_n = \sigma_h \cdot C_n$). In this section we will perform various combinations of symmetry operations and see whether or not the results we obtain represent equivalent orientations of the molecule we are studying. When we write the combination $A \cdot B$, the B operation will be performed first.

We have seen in the preceding section of this chapter that the NH_3 molecule has symmetry elements: E, C_3, and three σ_v. Because the principal axis of rotation has an order of 3, clockwise and counterclockwise operations with respect to this axis produce different results. Six symmetry operations are now available to us: the identity operation, a clockwise rotation, a counterclockwise rotation, and three reflections. If we look at possible combinations of these operations, we will find that the identity operation combined with any second operation will produce the equivalent orientation that the second operation alone would produce. The identity operation performed on any orientation of a molecule will not change that orientation. We will now try some combinations of operations that do not include the identity operation. For example, a $\overleftarrow{C}_3$ rotation followed by a σ_v reflection is shown in Figure 1.21. A counterclockwise rotation of 120° followed by a reflection through the vertical plane that includes hydrogen atom (2), or the lone reflection through the vertical plane that includes hydrogen atom (3), produces identical results [$\sigma_v(2) \cdot \overleftarrow{C}_3 = \sigma_v(3)$]. For a system of reference, the atoms of the molecule move with the operations,

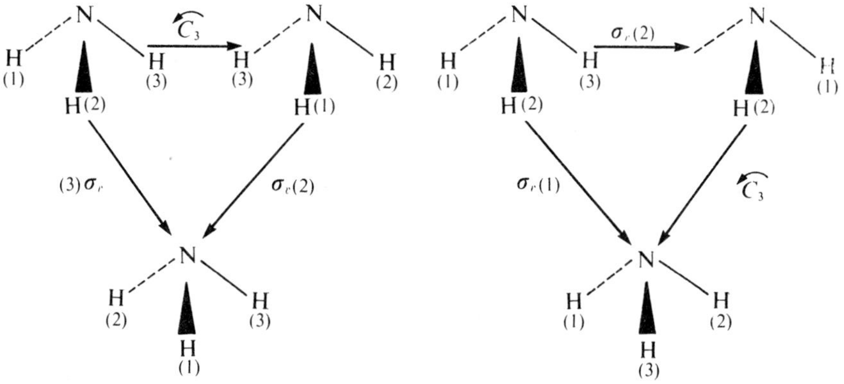

Figure 1.21
Combination of Symmetry Operations

but the number subscripts on the symmetry operations refer to the atoms as they were in the original orientation, in other words the symmetry elements remain fixed. Also as shown in Figure 1.21, $\overleftarrow{C_3} \cdot \sigma_v(2) = \sigma_v(1)$. As we will find upon examination, any combination in any order of the six symmetry operations present in the NH_3 molecule will produce results equivalent to one of the original operations. Another example is shown in Figure 1.22, where a clockwise rotation followed by a reflection with

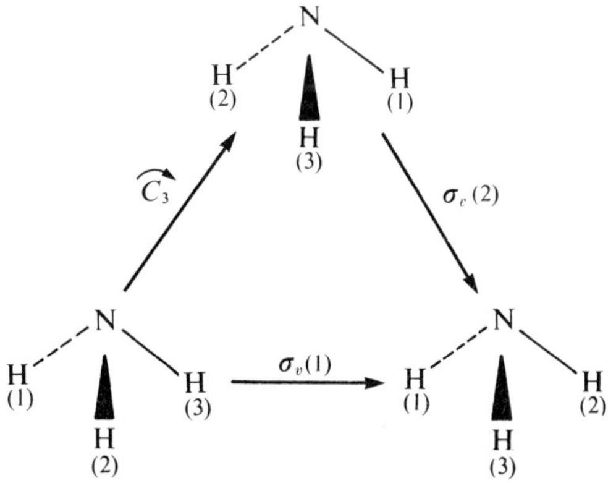

Figure 1.22
Combination of Symmetry Operations

respect to the $\sigma_v(2)$ plane produces the same orientation as reflection through the $\sigma_v(1)$ plane $[\sigma_v(2) \cdot \overrightarrow{C}_3 = \sigma_v(1)]$.

We will now construct a combination table for the symmetry operations present in the NH_3 molecule, in which the operation in the row is performed before the operation in the column. As seen in Figure 1.23, any combination of symmetry operations, or any operation performed twice, produces one of the six original operations. If this is true, then a third successive operation will also produce one of the original operations. The group of six symmetry operations previously mentioned are the only possible operations for the NH_3 molecule, because any combination of them produces the same results that one of the six original operations produces.

The molecule *trans*-$M(AB)_2X_2$, where AB represents a bidentate ligand that spans two adjacent positions, represents a compound with donor atoms roughly at vertices of an octahedran as shown in Figure 1.24. We will first identify the symmetry elements in this molecule. The identity element and a C_2 axis of rotation are easily recognized. No C_2 axes of rotation which are perpendicular to the principal C_2 axis are present, because a rotation about the B atoms or the A atoms shifts the position of the bidentate linkage. A horizontal plane of symmetry is the only symmetry plane present in this molecule for the same reason: any reflection across a vertical plane of symmetry would shift the bidentate linkage. A center of symmetry is located at the metal atom, but no improper axis of rotation (other than S_2) exists in this molecule. This generalized coordination compound contains only four symmetry elements: E, C_2, i, and σ_h.

If we investigate combinations of these operations, we will again find that no new operation is generated. In Figure 1.25 a rotation about the C_2 axis followed by a reflection through the horizontal plane of symmetry produces the same result as an inversion through the center of symmetry $(\sigma_h \cdot C_2 = i)$. The combination table for the symmetry operations contained in the $M(AB)_2X_2$ molecule is shown in Figure 1.26.

	E	$\overrightarrow{C}_3$	$\overleftarrow{C}_3$	$\sigma_v(1)$	$\sigma_v(2)$	$\sigma_v(3)$	(Row)
E	E	$\overrightarrow{C}_3$	$\overleftarrow{C}_3$	$\sigma_v(1)$	$\sigma_v(2)$	$\sigma_v(3)$	
$\overrightarrow{C}_3$	$\overrightarrow{C}_3$	$\overleftarrow{C}_3$	E	$\sigma_v(2)$	$\sigma_v(3)$	$\sigma_v(1)$	
$\overleftarrow{C}_3$	$\overleftarrow{C}_3$	E	$\overrightarrow{C}_3$	$\sigma_v(3)$	$\sigma_v(1)$	$\sigma_v(2)$	
$\sigma_v(1)$	$\sigma_v(1)$	$\sigma_v(3)$	$\sigma_v(2)$	E	$\overleftarrow{C}_3$	$\overrightarrow{C}_3$	
$\sigma_v(2)$	$\sigma_v(2)$	$\sigma_v(1)$	$\sigma_v(3)$	$\overrightarrow{C}_3$	E	$\overleftarrow{C}_3$	
$\sigma_v(3)$	$\sigma_v(3)$	$\sigma_v(2)$	$\sigma_v(1)$	$\overleftarrow{C}_3$	$\overrightarrow{C}_3$	E	

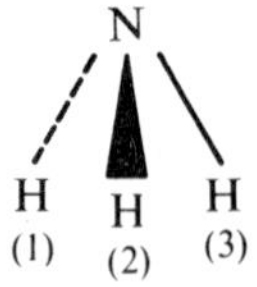

Figure 1.23
Combination Table For NH_3 (*Operations listed in the row are performed first*.)

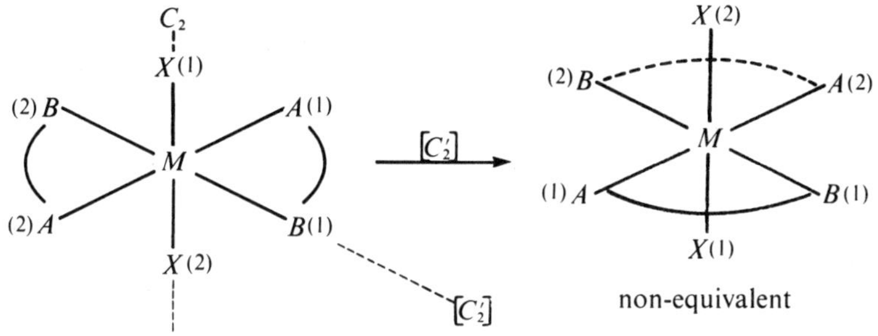

Figure 1.24
C_2 Rotations in $M(AB)_2X_2$

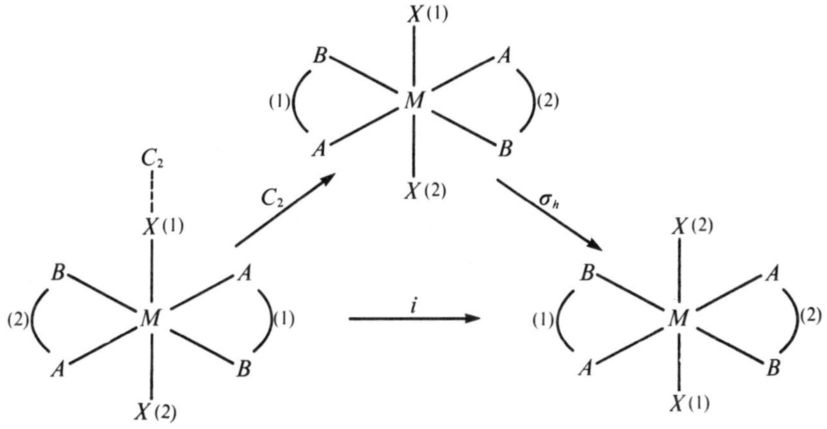

Figure 1.25
Combination of Symmetry Operations

	E	C_2	i	σ_h
E	E	C_2	i	σ_h
C_2	C_2	E	σ_h	i
i	i	σ_h	E	C_2
σ_h	σ_h	i	C_2	E

Figure 1.26
Combination Table For $M(AB)_2X_2$

Once all the symmetry elements with their respective operations are determined for a specific molecule, the combination of operations will never produce any additional symmetry operations. There are some extremely important consequences of this result, and a discussion of these consequences forms the basis for the next chapter.

Exercises

1–1: Identify all of the symmetry elements in the following molecules: H_2O, *cis*- $Pt(NH_3)_2Cl_2$, and BF_2Cl.

1–2: List all the operations that are associated with the symmetry elements found in Exercise 1–1.

1–3: Construct a combination table for the symmetry operations associated with the SF_5Cl molecule.

Bibliography

There are a number of very good books which are devoted to the use of symmetry and group theory in chemistry. Among these are

F. A. Cotton, *Chemical Applications of Group Theory*, 2d ed., New York: Wiley-Interscience, 1971.

R. M. Hochstrasser, *Molecular Aspects of Symmetry*, New York: W. A. Benjamin, Inc., 1966.

H. H. Jaffé and M. Orchin, *Symmetry in Chemistry*, New York: John Wiley and Sons, 1965.

D. S. Schonland, *Molecular Symmetry*, New York: D. Van Nostrand Company Ltd., 1965.

2

REPRESENTATIONS AND GROUPS

At the end of the last chapter we illustrated the systematic grouping of symmetry elements and symmetry operations by constructing combination tables. Now we would like to discuss the rules that define a mathematical group.

2-1 RULES OF GROUP THEORY

A *mathematical group* is a collection of members which are interrelated according to the following set of rules:

1. The combination of any two members of the group, including the square of each member, must yield a member of the group, i.e., $AB = C$, where A, B, and C are members of the group.
2. The group must contain an identity member E, where $EX = XE$. That is, E commutes with all members of the group.
3. The associative law of group combination must hold. This can be written in general terms as $X(YZ) = (XY)Z$.
4. Every member of the group, say A, must have a reciprocal, A^{-1}, which is also a member of the group such that $AA^{-1} = A^{-1}A = E$.

You may have already recognized the systematic grouping mentioned in Chapter 1 as the relationship that constitutes a mathematical group. For example, the symmetry operations present in the trans configuration of the $M(AB)_2X_2$ molecule were E, C_2, i, and σ_h. The combination table for these operations showed that no new operations were generated when any combination of these operations was taken; this illustrates rule (1). Because our operations included E, an identity operation, the second rule is obeyed. To test rule (3) we should perform a specific set of operations, such as $C_2(i\cdot\sigma_h)$ and $(C_2\cdot i)\sigma_h$, and see if we obtain identical results. Referring to the combination table in Figure 1.26, we obtain the following:

$$C_2\cdot(i\cdot\sigma_h) \stackrel{?}{=} (C_2\cdot i)\cdot\sigma_h$$

$$C_2\cdot C_2 \stackrel{?}{=} \sigma_h\cdot\sigma_h$$

$$E = E$$

Not only is rule (3) satisfied, but rule (4) is illustrated in the second and third steps of the above progression, since in this group each operation is its own reciprocal. In conclusion we note that the collection of all symmetry operations present in a molecule obeys the rules which characterize a mathematical group.

The *order* of a group, designated by h, is equal to the number of symmetry operations present in the group. For the $M(AB)_2X_2$ example, $h = 4$ i.e., (E, C_2, i, and σ_h).

Another concept important to the use of groups is *class*. By definition the *conjugate element* of A is XAX^{-1} where X and A are both members of the group. The members of a group which are conjugate to one another form a *class*. The order of a class is an *integral factor* of the order of the group.

As an example, we will construct the conjugate elements for the symmetry operations present in the NH_3 molecule (E, $\overrightarrow{C}_3$, $\overleftarrow{C}_3$, $\sigma_v(1)$, $\sigma_v(2)$, and $\sigma_v(3)$), for which the combination table in Figure 1.23 will prove useful. The identity operation, E, always forms a class by itself. If we perform the operations from right to left, we get $XEX^{-1} = XX^{-1} = E$. In Figure 2.1 the clockwise rotation about the C_3 axis is considered. By our definition $\overrightarrow{C}_3$ and $\overleftarrow{C}_3$ are conjugate elements and form a class with order equal to two. If we had chosen the counterclockwise rotation about the C_3 axis, we would have arrived at the same result. In Figure 2.2 the $\sigma_v(3)$ plane of symmetry is treated. In this case we find that $\sigma_v(1)$, $\sigma_v(2)$, and $\sigma_v(3)$ are conjugate elements and form a class of order three. We could have treated either $\sigma_v(1)$ or $\sigma_v(2)$ and found the same result. The six symmetry opera-

$$
\begin{aligned}
E\cdot\overrightarrow{C}_3\cdot E^{-1} &= E\cdot\overrightarrow{C}_3 = \overrightarrow{C}_3\\
\overrightarrow{C}_3\cdot\overrightarrow{C}_3\cdot\overrightarrow{C}_3^{-1} &= \overrightarrow{C}_3\cdot E = \overrightarrow{C}_3\\
\overleftarrow{C}_3\cdot\overrightarrow{C}_3\cdot\overleftarrow{C}_3^{-1} &= \overleftarrow{C}_3\cdot\overleftarrow{C}_3 = \overrightarrow{C}_3\\
\sigma_v(1)\cdot\overrightarrow{C}_3\cdot\sigma_v(1)^{-1} &= \sigma_v(1)\cdot\sigma_v(2) = \overleftarrow{C}_3\\
\sigma_v(2)\cdot\overrightarrow{C}_3\cdot\sigma_v(2)^{-1} &= \sigma_v(2)\cdot\sigma_v(3) = \overleftarrow{C}_3\\
\sigma_v(3)\cdot\overrightarrow{C}_3\cdot\sigma_v(3)^{-1} &= \sigma_v(3)\cdot\sigma_v(1) = \overleftarrow{C}_3
\end{aligned}
$$

Figure 2.1
C_3 Conjugate Elements

$$
\begin{aligned}
E\cdot\sigma_v(3)\cdot E^{-1} &= E\cdot\sigma_v(3) = \sigma_v(3)\\
\overrightarrow{C}_3\cdot\sigma_v(3)\cdot\overrightarrow{C}_3^{-1} &= \overrightarrow{C}_3\cdot\sigma_v(1) = \sigma_v(2)\\
\overleftarrow{C}_3\cdot\sigma_v(3)\cdot\overleftarrow{C}_3^{-1} &= \overleftarrow{C}_3\cdot\sigma_v(2) = \sigma_v(1)\\
\sigma_v(1)\cdot\sigma_v(3)\cdot\sigma_v(1)^{-1} &= \sigma_v(1)\cdot\overleftarrow{C}_3 = \sigma_v(2)\\
\sigma_v(2)\cdot\sigma_v(3)\cdot\sigma_v(2)^{-1} &= \sigma_v(2)\cdot\overrightarrow{C}_3 = \sigma_v(1)\\
\sigma_v(3)\cdot\sigma_v(3)\cdot\sigma_v(3)^{-1} &= \sigma_v(3)\cdot E = \sigma_v(3)
\end{aligned}
$$

Figure 2.2
σ_v Conjugate Elements

tions associated with the NH_3 molecule fall into three classes: identity, rotations, and reflections. The orders of these three classes are one, two, and three, respectively; it should be noted that these values are all integral factors of the order of the group, and that the sum of the orders of classes equals the order of the group.

The systematic grouping of symmetry operations that we have discussed is called a *point group* and has all the properties of a mathematical group. There are two types of notation used by chemists to designate point groups. These are the Schoenflies system, which is favored by spectroscopists, and the Hermann-Mauguin or international system, which is favored by crystallographers and others interested in the properties of the crystalline state. In this chapter we will describe the Schoenflies system and defer the description of the international system until Chapter 8. In both these systems the symbol for the point group summarizes the essential symmetry operations in the group, or at least is a mnemonic device for the point group.

Consider the simplest of all possible groups; that being the group constituted by the element E only. Since $E = C_1$, this group is assigned the

Schoenflies symbol C_1. Logically, then, those groups which are made up of rotation operations only, about the same major axis, are designated C_n, where n is the order of the axis. For example, for the group made up by E and C_2, the Schoenflies symbol is C_2, and for the group made up by E, C_3^1, C_3^2 the symbol is C_3.*

The group constituted by the elements E and i is assigned the symbol C_i while that group which includes only the identity element and a mirror plane is C_s.

Next we consider those groups which have, in addition to E and C_n, a set of σ_v planes. A little reflection on this situation will reveal that if there are σ_v planes then there must be n of these planes. The point groups are given the symbol C_{nv}. The water molecule which possesses E, C_2 and $2\sigma_v$ planes (which generate four operations) belongs to the point group C_{2v}, while ammonia, which has E, C_3 and $3\sigma^k$ (which generate E, C_3^1, C_3^2, σ_v^a, σ_v^b, σ_v^c operations) belongs to the point group C_{3v}. It is obvious that there are an infinite number of C_{nv} groups, but only a limited number of these are of chemical importance.

It is easy to visualize situations in which there are C_2 axes perpendicular to the major axis (we encountered several in Chapter 1). These groups are given the symbol D_n, D_{nh} or D_{nd} depending on the presence and orientation of the mirror planes. The $PtCl_4^{2-}$ molecule, which has σ_h, C_4 and $4C_2$ axes perpendicular to the C_4 axis, belongs to the group D_{4h}, while BF_3 belongs to the group D_{3h}.

Exercise

2-1: Write out the symmetry elements and the complete set of corresponding symmetry operations for the following molecules: $PtCl_4{}^{2-}$, C_6H_6 (benzene), BF_3, trigonal bipyramidal PF_5, tetragonal pyramidal $Ni(CN)_5^{3-}$, trans–$[Co(NH_3)_4Cl_2]^+$ (consider free rotation about the Co–N bonds).

In addition to a number of other types of point groups with which we will become familiar in later chapters, it is necessary at this point to call attention to three additional point groups, and these are constituted by the operations found in octahedral, tetrahedral, and icosahedral molecules. The symbols are O_h, T_d, and I_h, respectively.

---

*In order to emphasize the direction of the rotations, we have used elsewhere the symbols $\overrightarrow{C}_3$ and $\overleftarrow{C}_3$ to indicate the rotation operations. It should be clear that $\overleftarrow{C}_3 = C_3^1$, $\overrightarrow{C}_3 = C_3^2$, $\overleftarrow{C}_4 = C_4^1$, and $\overrightarrow{C}_4 = C_4^3$.

Exercise

2–2: List the symmetry elements and operations for tetrahedral, octahedral, and icosahedral objects.

We will now outline the procedure used for classifying molecules into these point groups. As soon as we know to what point group a molecule belongs, the symmetry operations present in that molecule are determined, because each point group contains a specific set of symmetry operations.

The first step is to determine whether or not a molecule belongs to one of the special point groups, $C_{\infty v}$, $D_{\infty h}$, O_h (octahedral), T_d (tetrahedral), or I_h (icosahedral). Only a linear molecule can belong to one of the first two groups. If a molecule appears to belong to any of these special groups, the required symmetry operations should still be verified. Rather than proceeding with a number of rules for assignment of point groups, a flow diagram for classifying molecules into point groups is presented in Figure 2.3.

The NH_3 molecule is one that we have considered in detail already. We found the symmetry elements to be E, two C_3's, and three σ_v's. The NH_3 molecule is not linear nor does it belong to one of the other special point groups, so we continue through the flow diagram to the point where we must look for proper axes of rotation. If the molecule in question contains more than one proper axis of rotation, we select the one of highest order. A C_3 axis in NH_3 is the proper axis of highest order. There is no S_6 axis of rotation nor are there any C_2 axes which are perpendicular to the principal axis, C_3, so S groups and D groups are eliminated. There is no σ_h plane of symmetry, but there are σ_v planes present. The NH_3 molecule must, therefore, belong to the C_{3v} point group. Every molecule, that belongs to C_{3v}, not just NH_3, must contain E, two C_3's, and three σ_v's as symmetry operations.

The PF_5 molecule, which is shown in Figure 2.4, has a trigonal bipyramidal structure. This molecule was found to contain E, two C_3's, three C_2's, σ_h, two S_3's, and three σ_v's as symmetry elements. To determine to what point group the PF_5 molecule belongs, we will proceed through the flow diagram once again. The PF_5 molecule is not linear nor does it belong to any of the other special point groups. The proper rotation axis of highest order is the C_3 axis. There is no S_6 axis of rotation but there is a set of C_2 axes perpendicular to the C_3 axis; this means that PF_5 must belong to one of the D point groups. Because this molecule contains a σ_h plane of symmetry, the point group for PF_5 must be D_{3h}. Once again any molecule that falls into the D_{3h} classification must contain the symmetry operations which are included in this group.

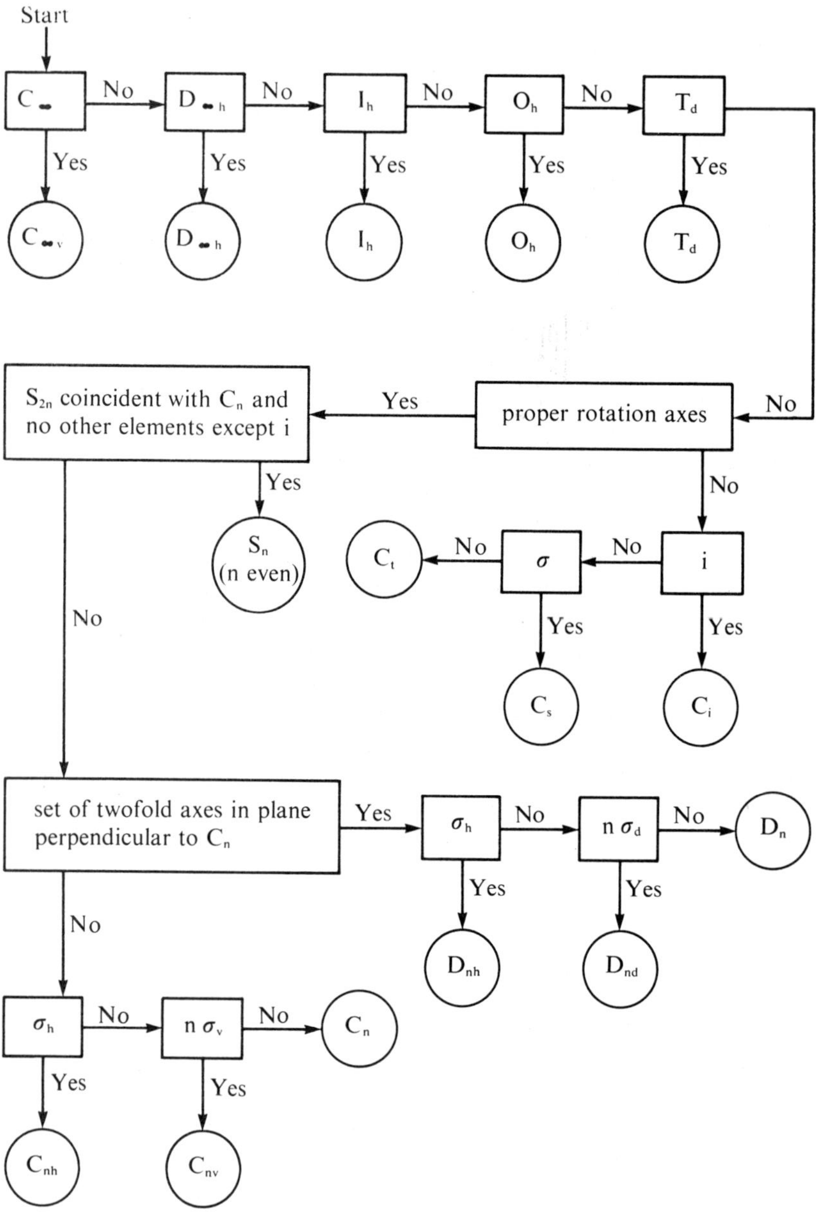

Figure 2.3
Flow Chart for Classifying Molecules into Point Groups

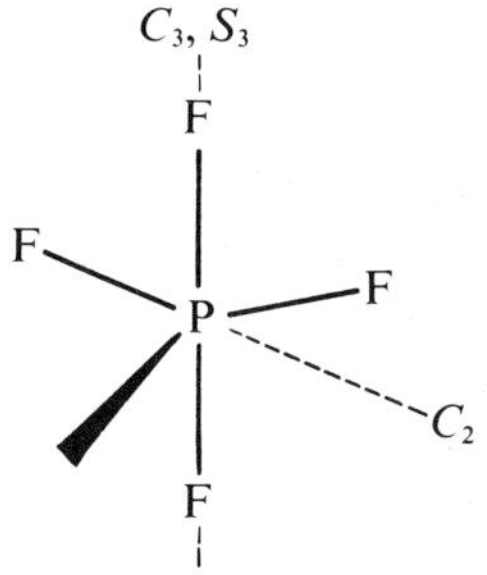

Figure 2.4
Rotation Axes of PF_5

In Exercise 1–3, Chapter 1, we constructed a combination table for the SF_5Cl molecule. As shown in Figure 2.5 this molecule has an octahedral shape, but does not fall in the O_h point group because all six positions of the octahedron are not equivalent. The symmetry elements present in the SF_5Cl molecule are E, C_4, C_2, two σ_v's and two σ_d's. Let us now see what point group is composed by the symmetry operations generated by these symmetry elements, or in other words, to which point group does SF_5Cl belong. This molecule is not linear nor does it belong to any of the other special groups. The proper rotation axis of highest order is a C_4 axis. No S_8 axis and no C_2 axes which are perpendicular to the C_4 axis are present. There is also no horizontal plane of symmetry, but there are vertical planes of symmetry, and this means that the molecule belongs to the point group C_{4v}.

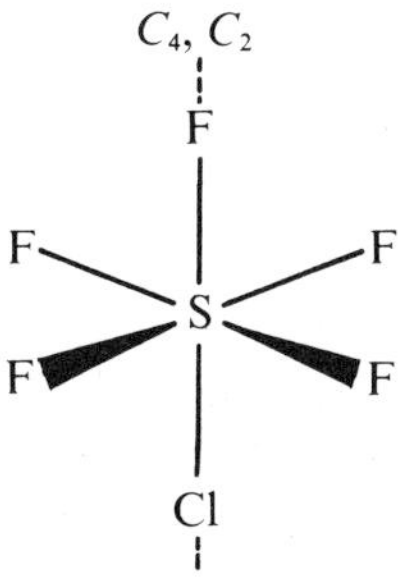

Figure 2.5
Rotation Axes of SF_5Cl

Exercises

2–3: Assign the molecules in Exercise 2–1 to the appropriate point group i.e., give the Schoenflies symbol.
2–4: Determine the class structure of the symmetry operations in the D_{3h} point group (PF_5) and the C_{4v} point group (SF_5Cl).
2–5: Classify the following molecules into point groups: H_2O, BF_3, and trans–$Pt(NH_3)_2Cl_2$.

2-2 REPRESENTATION THEORY

A matrix provides a way to mathematically represent a movement of an object through space. In this section we will illustrate that a symmetry operation, or the associated molecular movement, can best be represented by a square matrix. A square matrix has the same number of rows and columns, and the size of our square matrices will depend on the dimensionality of the basis set that is selected.

If a square matrix M_A, M_B, M_C, . . . can be associated with every operation of a group $A, B, C, \ldots$ such that, for example, if $AB = C$, then $M_A M_B = M_C$, then the set of matrices form a group which is identical with the group $A, B, C, \ldots$. Such matrices are a representation of the group, and their order is the same as that of the group.

Before proceeding with the representations, a few examples of matrix multiplication will be demonstrated. In Equation (2–1) two second order

$$\begin{pmatrix} A & B \\ C & D \end{pmatrix} \begin{pmatrix} X & Y \\ W & Z \end{pmatrix} = \begin{pmatrix} AX + BW & AY + BZ \\ CX + DW & CY + DZ \end{pmatrix} \qquad \textbf{(2-1)}$$

square matrices are multiplied. The procedure followed is to multiply a row in the left matrix times a column in the right matrix; each entry in the product matrix is a summation of the term by term products. Therefore, the number of columns in the first (left) matrix must equal the number of rows in the second matrix. For example, in Equation (2–2) a 2×2 matrix is multiplied times a 2×1 matrix. There are two rows and two columns in the matrix on the left and two rows and one column in the second matrix.

$$\begin{pmatrix} 3 & 1 \\ 2 & 4 \end{pmatrix} \begin{pmatrix} 2 \\ 7 \end{pmatrix} = \begin{pmatrix} 6 + 7 \\ 4 + 28 \end{pmatrix} = \begin{pmatrix} 13 \\ 32 \end{pmatrix} \qquad \textbf{(2-2)}$$

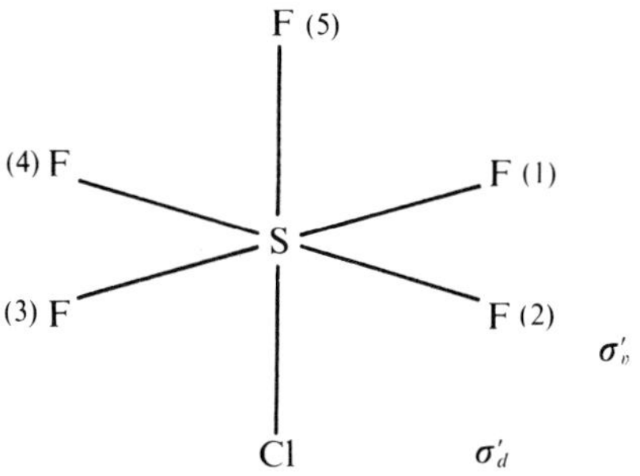

	E	$\overrightarrow{C_4}$	$\overleftarrow{C_4}$	C_2	σ_v	σ_v'	σ_d	σ_d'
E	E	$\overrightarrow{C_4}$	$\overleftarrow{C_4}$	C_2	σ_v	σ_v'	σ_d	σ_d'
$\overrightarrow{C_4}$	$\overrightarrow{C_4}$	C_2	E	$\overleftarrow{C_4}$	σ_d	σ_d'	σ_v'	σ_v
$\overleftarrow{C_4}$	$\overleftarrow{C_4}$	E	C_2	$\overrightarrow{C_4}$	σ_d'	σ_d	σ_v	σ_v'
C_2	C_2	$\overleftarrow{C_4}$	$\overrightarrow{C_4}$	E	σ_v'	σ_v	σ_d'	σ_d
σ_v	σ_v	σ_d'	σ_d	σ_v'	E	C_2	$\overleftarrow{C_4}$	$\overrightarrow{C_4}$
σ_v'	σ_v'	σ_d	σ_d'	σ_v	C_2	E	$\overrightarrow{C_4}$	$\overleftarrow{C_4}$
σ_d	σ_d	σ_v	σ_v'	σ_d'	$\overrightarrow{C_4}$	$\overleftarrow{C_4}$	E	C_2
σ_d'	σ_d'	σ_v'	σ_v	σ_d	$\overleftarrow{C_4}$	$\overrightarrow{C_4}$	C_2	E

Figure 2.6
Combination Table For SF_5Cl

We will first examine the SF_5Cl molecule which belongs to the C_{4v} point group. The combination table for this group is shown in Figure 2.6.

Because we are dealing with three dimensional space we will construct one matrix which represents a three dimensional coordinate system,

$$\begin{pmatrix} x \\ y \\ z \end{pmatrix}$$

Now we must find what 3×3 matrices represent the various symmetry operations in the C_{4v} point group. In Equation (2-3) the matrix

$$\begin{pmatrix} 1 & 0 & 0 \\ 0 & 1 & 0 \\ 0 & 0 & 1 \end{pmatrix}$$

is shown to represent the identity operation; when this matrix is multiplied times the coordinate matrix, the orientation of the coordinate system of the molecule has not changed, which is precisely what is meant by an identity operation.

$$E\begin{pmatrix} x \\ y \\ z \end{pmatrix} = \begin{pmatrix} 1 & 0 & 0 \\ 0 & 1 & 0 \\ 0 & 0 & 1 \end{pmatrix}\begin{pmatrix} x \\ y \\ z \end{pmatrix} = \begin{pmatrix} x \\ y \\ z \end{pmatrix} \qquad \textbf{(2-3)}$$

In Figure 2.7 a $\overrightarrow{C_4}$ operation can be seen to move the y-axis into the x-axis and the x-axis into the negative y direction. The z-axis, which is

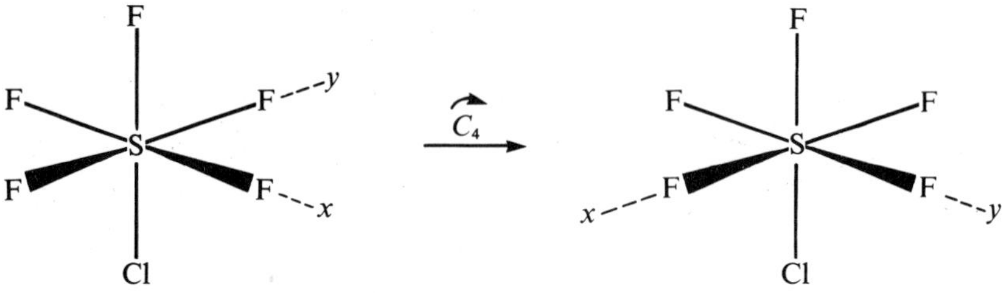

Figure 2.7
The C_4 Operation

colinear with the C_4 axis, is unchanged. In Equation (2-4) the clockwise

$$\vec{C}_4\begin{pmatrix}x\\y\\z\end{pmatrix}=\begin{pmatrix}y\\-x\\z\end{pmatrix}=\begin{pmatrix}0&1&0\\-1&0&0\\0&0&1\end{pmatrix}\begin{pmatrix}x\\y\\z\end{pmatrix} \tag{2-4}$$

rotation about the C_4 axis of rotation is examined. The 3×3 matrix illustrated there represents the $\vec{C}_4$ symmetry operation. Equation (2-5) illustrates the matrix representation for a σ_v symmetry operation

$$\sigma_v\begin{pmatrix}x\\y\\z\end{pmatrix}=\begin{pmatrix}1&0&0\\0&-1&0\\0&0&1\end{pmatrix}\begin{pmatrix}x\\y\\z\end{pmatrix} \tag{2-5}$$

in the C_{4v} point group. The σ_v symmetry plane that is illustrated is the xz plane; therefore, the only change is that of the y-coordinate changing sign by a reflection across the plane of symmetry. Figure 2.8 shows the matrix representations for all the symmetry operations in the C_{4v} point group.

Our next objective is to illustrate that any combination of these matrices produces a representation matrix within the group. For example, from the combination table in Figure 2.6 we see that $\sigma_v \cdot \vec{C}_4 = \sigma_{d'}$. Remembering that we perform the $\vec{C}_4$ operation first, we now show in Equation (2-6) the

$$\underset{\sigma_v}{\begin{pmatrix}1&0&0\\0&-1&0\\0&0&1\end{pmatrix}}\underset{\vec{C}_4}{\begin{pmatrix}0&1&0\\-1&0&0\\0&0&1\end{pmatrix}}=\underset{\sigma_{d'}}{\begin{pmatrix}0&1&0\\1&0&0\\0&0&1\end{pmatrix}} \tag{2-6}$$

$$E\begin{pmatrix} x \\ y \\ z \end{pmatrix} = \begin{pmatrix} 1 & 0 & 0 \\ 0 & 1 & 0 \\ 0 & 0 & 1 \end{pmatrix}\begin{pmatrix} x \\ y \\ z \end{pmatrix} = \begin{pmatrix} x \\ y \\ z \end{pmatrix}$$

$$\overrightarrow{C}_4\begin{pmatrix} x \\ y \\ z \end{pmatrix} = \begin{pmatrix} 0 & 1 & 0 \\ -1 & 0 & 0 \\ 0 & 0 & 1 \end{pmatrix}\begin{pmatrix} x \\ y \\ z \end{pmatrix} = \begin{pmatrix} y \\ x \\ z \end{pmatrix}$$

$$\overleftarrow{C}_4\begin{pmatrix} x \\ y \\ z \end{pmatrix} = \begin{pmatrix} 0 & -1 & 0 \\ 1 & 0 & 0 \\ 0 & 0 & 1 \end{pmatrix}\begin{pmatrix} x \\ y \\ z \end{pmatrix} = \begin{pmatrix} -y \\ x \\ z \end{pmatrix}$$

$$C_2\begin{pmatrix} x \\ y \\ z \end{pmatrix} = \begin{pmatrix} -1 & 0 & 0 \\ 0 & -1 & 0 \\ 0 & 0 & 1 \end{pmatrix}\begin{pmatrix} x \\ y \\ z \end{pmatrix} = \begin{pmatrix} -x \\ -y \\ z \end{pmatrix}$$

$$\sigma_v\begin{pmatrix} x \\ y \\ z \end{pmatrix} = \begin{pmatrix} 1 & 0 & 0 \\ 0 & -1 & 0 \\ 0 & 0 & 1 \end{pmatrix}\begin{pmatrix} x \\ y \\ z \end{pmatrix} = \begin{pmatrix} x \\ -y \\ z \end{pmatrix}$$

$$\sigma_v'\begin{pmatrix} x \\ y \\ z \end{pmatrix} = \begin{pmatrix} -1 & 0 & 0 \\ 0 & 1 & 0 \\ 0 & 0 & 1 \end{pmatrix}\begin{pmatrix} x \\ y \\ z \end{pmatrix} = \begin{pmatrix} -x \\ y \\ z \end{pmatrix}$$

$$\sigma_d\begin{pmatrix} x \\ y \\ z \end{pmatrix} = \begin{pmatrix} 0 & -1 & 0 \\ -1 & 0 & 0 \\ 0 & 0 & 1 \end{pmatrix}\begin{pmatrix} x \\ y \\ z \end{pmatrix} = \begin{pmatrix} -y \\ -x \\ z \end{pmatrix}$$

$$\sigma_d'\begin{pmatrix} x \\ y \\ z \end{pmatrix} = \begin{pmatrix} 0 & 1 & 0 \\ 1 & 0 & 0 \\ 0 & 0 & 1 \end{pmatrix}\begin{pmatrix} x \\ y \\ z \end{pmatrix} = \begin{pmatrix} y \\ x \\ z \end{pmatrix}$$

Figure 2.8
Matrix Representations of C_{4v}

product of the two matrices which represent these symmetry operations. The matrix representations give the same result as the symmetry operations themselves; this is true for any combination of representation matrices.

Before we illustrate that the set of matrices representing the symmetry operations in the C_{4v} point group belongs to a mathematical group, we must first demonstrate another matrix operation called the inversion operation. The inverse of the matrix which represents the $\overleftarrow{C}_4$ operation will be constructed first. In Figure 2.9 the $\overleftarrow{C}_4$ matrix along with its transpose

$$\begin{pmatrix} 0 & -1 & 0 \\ 1 & 0 & 0 \\ 0 & 0 & 1 \end{pmatrix} \qquad \begin{pmatrix} 0 & 1 & 0 \\ -1 & 0 & 0 \\ 0 & 0 & 1 \end{pmatrix} \qquad \begin{pmatrix} 0 & 1 & 0 \\ -1 & 0 & 0 \\ 0 & 0 & 1 \end{pmatrix}$$

$\overleftarrow{C}_4$ transpose matrix co-factor matrix

Figure 2.9
Matrices For $\overleftarrow{C}_4$

matrix and the co-factor matrix of the transpose matrix are presented. The transpose matrix is a matrix that results from interchanging the rows and columns of the original matrix, or in other words, if a_{ij} represented the entries in the original matrix, then a_{ji} represents the entries in its transpose matrix. The co-factor of any entry, a_{ij}, is $(-1)^{i+j}$ times the determinant of the matrix obtained by deleting the i th row and j th column of the matrix. In Equation (2-7) two terms of the co-factor matrix are worked out

$$\begin{aligned} &\begin{pmatrix} 0 & 1 & 0 \\ 1 & 0 & 0 \\ 0 & 0 & 1 \end{pmatrix} \quad a_{11} = (-1)^2[(0 \times 1) - (0 \times 0)] = 0 \\ &\begin{pmatrix} 0 & 1 & 0 \\ 1 & 0 & 0 \\ 0 & 0 & 1 \end{pmatrix} \quad a_{21} = (-1)^3[(1 \times 1) - (0 \times 0)] = -1 \end{aligned} \tag{2-7}$$

for the transpose matrix in Figure 2.9. The other terms of the co-factor matrix can be found in a similar manner. The value of the determinant of a matrix is found by diagonal multiplication of the entries. A 2×2 determinant and a 3×3 determinant are illustrated in Figure 2.10. The

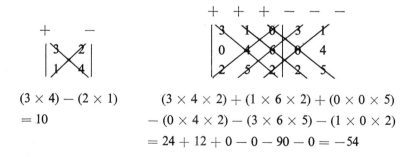

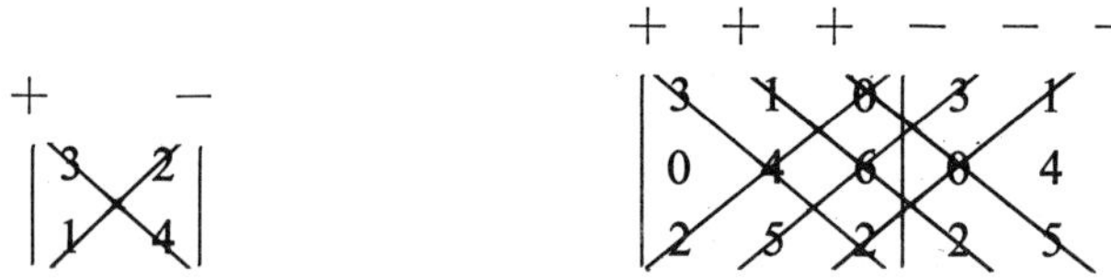

$(3 \times 4) - (2 \times 1)$
$= 10$

$(3 \times 4 \times 2) + (1 \times 6 \times 2) + (0 \times 0 \times 5)$
$- (0 \times 4 \times 2) - (3 \times 6 \times 5) - (1 \times 0 \times 2)$
$= 24 + 12 + 0 - 0 - 90 - 0 = -54$

Figure 2.10
Evaluation of Two-by-Two and Three-by-Three Determinants

diagonal terms that are multiplied from left to right are summed and the products of the terms in diagonals from right to left are subtracted. A 4×4 determinant, or larger, cannot be expanded in this manner. For the procedure see a textbook on mathematics.

We are finally ready to calculate the inverse matrix for the representation of the $\overleftarrow{C}_4$ operation. This matrix is found by dividing the co-factor matrix of the transpose matrix by the value of the determinant of the original matrix that represents the $\overleftarrow{C}_4$ operation. We shall construct the inverse in steps. First the cofactor matrix Equation (2-8) is divided by

$$\begin{pmatrix} 0 & 1 & 0 \\ -1 & 0 & 0 \\ 0 & 0 & 1 \end{pmatrix} \qquad \textbf{(2-8)}$$

the determinant of the original matrix, which, as shown in Equation (2-9)

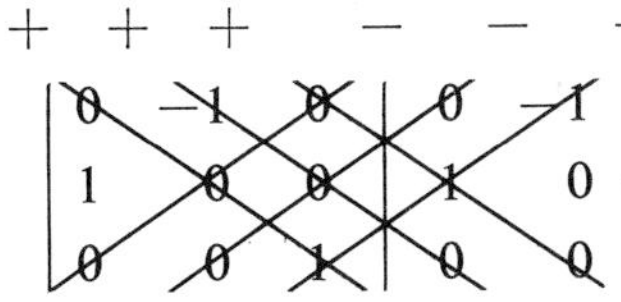

$$= 0 + 0 + 0 - 0 - 0 - (-1) = 1 \qquad \textbf{(2-9)}$$

is equal to 1. (In dividing a matrix by a constant, each element in the matrix is divided by that constant.) As a check, take the product of the original representation matrix for $\overleftarrow{C}_4$ and its inverse, which we have just constructed, since, according to the fourth rule, $\overleftarrow{C}_4 \cdot (\overleftarrow{C}_4)^{-1} = E$. The product is Equation (2-10).

$$\underset{\overleftarrow{C}_4}{\begin{pmatrix} 0 & -1 & 0 \\ 1 & 0 & 0 \\ 0 & 0 & 1 \end{pmatrix}} \underset{(\overleftarrow{C}_4)^{-1}}{\begin{pmatrix} 0 & 1 & 0 \\ -1 & 0 & 0 \\ 0 & 0 & 1 \end{pmatrix}} = \underset{E}{\begin{pmatrix} 1 & 0 & 0 \\ 0 & 1 & 0 \\ 0 & 0 & 1 \end{pmatrix}} \qquad \textbf{(2-10)}$$

We now point out that the matrix representing $(\overleftarrow{C}_4)^{-1}$ is the same as the matrix which represents $\overrightarrow{C}_4$, as it should be since $\overleftarrow{C}_4 \cdot \overrightarrow{C}_4 = E$.

If we now recall the four rules that must be obeyed for matrices to qualify as a mathematical group, we may check to see if our matrix representations of symmetry operations constitute a group. In Equation (2-6) we have already shown that the combination of any two members of the group yields a member of the group. That the combination of any two members of the group yields another member may be proven by an exhaustive process. We leave this as an exercise for the reader. An identity matrix has been illustrated, so that the second rule is obeyed. The associative law of multiplication holds; this is also left as an exercise for the reader. Every representation matrix has a reciprocal, or inverse, as we have just shown for one case. Therefore, the matrices that represent the symmetry operations of a point group belong to a group that is isomorphic with the group of symmetry operations.

Reducible and irreducible representations are two concepts which we should now discuss. Sets of representation matrices that can be block diagonalized are *reducible representations* and those which cannot be reduced in this manner are *irreducible representations*. For example, if we examine the matrix representations for the C_{4v} point group in Figure 2.8, the x and y coordinates are never mixed with the z coordinate by the performance of the operations. As they are written, this set of representation matrices can be simplified or reduced. Two examples are shown in Figure 2.11. The

$$\text{trace} = 1 \quad \begin{pmatrix} 1 & 0 & 0 \\ 0 & -1 & 0 \\ 0 & 0 & 1 \end{pmatrix}; \quad \begin{pmatrix} 1 & 0 \\ 0 & -1 \end{pmatrix} \quad \begin{matrix} \text{trace} = 0 \\ \\ (1) \qquad \text{trace} = 1 \end{matrix}$$

$$\sigma_v$$

$$\text{trace} = 1 \quad \begin{pmatrix} 0 & 1 & 0 \\ -1 & 0 & 0 \\ 0 & 0 & 1 \end{pmatrix}; \quad \begin{pmatrix} 0 & 1 \\ -1 & 0 \end{pmatrix} \quad \begin{matrix} \text{trace} = 0 \\ \\ (1) \qquad \text{trace} = 1 \end{matrix}$$

$$\vec{C}_4$$

Figure 2.11
Reduction of Representation Matrices

character of any matrix is equal to the trace of that matrix, where the trace of a matrix is defined as the sum of the entries along the diagonal. Class has been defined earlier in this chapter; the representation matrices for symmetry operations in the same class must have equal traces or characters. Irreducible representations of a group are of fundamental importance, and their uses and properties will be described in the next section along with the organization of the characters of the representation matrices into character table form.

2-3 CHARACTER TABLES

Much of the information about molecular symmetry that we have discussed can be consolidated into what are known as character tables. These tables are constructed from the traces of the irreducible representation matrices for each class within a point group. Every point group has a corresponding character table.

Before we generate a character table there are four important rules that should be mentioned concerning irreducible representations and their characters.

1. The sum of the squares of the dimensions of the irreducible representations of the group is equal to the order of the group.

$$d_1^2 + d_2^2 + \cdots + d_n^2 = h$$

2. The sum of the squares of the characters in any irreducible representation is equal to the order of the group.

$$\chi_1^2 + \chi_2^2 + \cdots + \chi_n^2 = h$$

3. Any two irreducible representations are orthogonal, *i.e.*, the sum of the products of the characters representing each operation is equal to zero.
4. The number of irreducible representations of a group is equal to the number of classes in the group.

The best method for explaining these rules is to work through an example and generate the character table for a specific point group. The first example that we will consider is the trans–$M(AB)_2X_2$ coordination compound which belongs to the C_{2h} point group. As discussed in Chapter 1 this point group has four symmetry operations: E, C_2, i, and σ_h. Each of these operations constitutes a class. Each irreducible representation that is contained in the table must have four terms, or characters, and there must be four irreducible representations (rule 4). Because there are four operations in the C_{2h} point group, the order of the group is four, and according to rule 1, each irreducible representation must be of order 1, viz.,

$$d_1^2 + d_2^2 + d_3^2 + d_4^2 = h = 4$$

where the d's represent the dimensions of the irreducible representations. The only solution for this equation is that all the dimensions equal one. In Figure 2.12 we illustrate what has been determined to this point. If rule 2 is considered, we will discover that all the characters in the irreducible representations must be equal to $+1$ or -1, since there are four characters in each representation and the sum of their squares must equal four. Let us choose as our first irreducible representation four $+1$'s ($\Gamma_1 = 1\ 1\ 1\ 1$). In choosing the second irreducible representation we must follow rule 3, so that Γ_1 is orthogonal to Γ_2. The definition of orthogonality is that after

		E	C_2	i	σ_h ← operations
	Γ_1	1			
irreducible	Γ_2	1			
representations	Γ_3	1 ↖			
	Γ_4	1			
		dimensions			

Figure 2.12
Character Table Definitions

term by term multiplication, the sum of the products must equal zero. For example, if we let $\Gamma_2 = 1 \;\; 1 \;\; -1 \;\; -1$ and multiply this by Γ_1 thereby obtaining $[(1)(1) + (1)(1) + (-1)(1) + (-1)(1) = 0]$, we find that the sum of the products equals zero so Γ_1 and Γ_2 are orthogonal. In order to obey the rules we must use $+1$'s and -1's for the characters. Using these we find all four irreducible representations as shown in Figure 2.13. As a last check we find that all four rules are obeyed for this representation table.

	E	C_2	i	σ_h
Γ_1	1	1	1	1
Γ_2	1	1	-1	-1
Γ_3	1	-1	-1	1
Γ_4	1	-1	1	-1

Figure 2.13
C_{2h} Irreducible Representations

As a second example we will generate the C_{3v} character table. The NH_3 molecule belongs to this point group, which contains six symmetry operations: E, two C_3's, and three σ_v's. Earlier in this chapter we found that the six operations fell into three classes: E, $2C_3$, and $3\sigma_v$. Therefore, according to rule 4 there must be three irreducible representations in the C_{3v} character table. Rule 1 states that the sum of the squares of the dimensions of these three representations must equal the order of the group: $d_1^2 + d_2^2 + d_3^2 = 6$. The only values of d that will satisfy this requirement are 1, 1, and 2. Figure 2.14 illustrates our results to this point. Once again, in this and any other character table, there will be a one-dimensional irreducible representation that has all characters equal to $+1$. When checking to see if this representation (Γ_1) obeys rule 2, we must keep in mind that there are two axes of rotation and three planes of symmetry and each of these has a character of $+1$. Therefore, $1^2 + 2(1)^2 + 3(1)^2 = 6$ and rule 2 is obeyed.

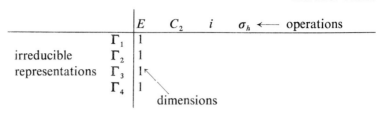

	E	$2C_3$	$3\sigma_v$
Γ_1	1		
Γ_2	1		
Γ_3	2		

Figure 2.14
Dimensions of C_{3v}

In seeking a representation that has all $+1$'s and -1's and is orthogonal to Γ_1, we must have three $+1$'s and three -1's, and since all operations in the same class must have equal characters, Γ_2 must equal 1 1 -1. The third representation is two-dimensional and in order for rule 2 to be obeyed the second character must be ± 1 ($\Gamma_3 = 2 \ \pm 1 \ 0$) so that the sum of the squares of all the characters equals 6. All three representations must be orthogonal to one another, so if we choose 2 $+1$ 0 for Γ_3 and multiply it times Γ_1 we have $1(1)(2) + 2(1)(1) + 3(1)(0) = 4$ and Γ_1 and Γ_3 are not orthogonal. Therefore, Γ_3 must equal 2 -1 0. The product of the correct Γ_3 and Γ_1 is $1(1)(2) + 2(1)(-1) + 3(1)(0) = 0$. The irreducible representations for the C_{3v} point group are given in Figure 2.15. It is left to the reader to show that Γ_1 and Γ_2, and Γ_2 and Γ_3 are orthogonal pairs.

	E	$2C_3$	$3\sigma_v$
Γ_1	1	1	1
Γ_2	1	1	-1
Γ_3	2	-1	0

Figure 2.15
C_{3v} Irreducible Representations

A complete character table for the C_{2h} point group as it appears in Appendix I is given in Figure 2.16. The point group to which a character table belongs is written in the upper left-hand corner. Instead of designat-

C_{2h}	E	C_2	i	σ_h		
A_g	1	1	1	1	R_z	x^2, y^2, z^2, xy
B_g	1	-1	1	-1	R_x, R_y	xz, yz
A_u	1	1	-1	-1	z	
B_u	1	-1	-1	1	x, y	

Figure 2.16
C_{2h} Character Table

ing the irreducible representations as $\Gamma_1, \Gamma_2, \ldots$ we now have A's, B's, E's, and T's which are the Mulliken symbols. All one-dimensional representations are designated A or B, all two-dimensional representations are designated E, and all three-dimensional representations are designated T. For the one dimensional irreducible representation the symbol is A if the character for rotation about the major axis is $+1$ and the rotation is symmetric, and the symbol is B if the character for rotation is -1 and the rotation is antisymmetric. A subscript of 1 on an A or B means that the irreducible representation is symmetric with respect to a C_2 perpendicular to the major axis, while a subscript of 2 means antisymmetric with respect to this operation. If there are no C_2 axes, then a vertical mirror plane is selected. Primes and double primes attached to all symbols mean symmetric and antisymmetric respectively, with respect to σ_h, while g and u subscripts refer to the inversion operation. If the character under i is $+1$, the g subscript is used, and the u subscript is used if the character under i is -1.

In the two right hand columns of the character table, there appear several symbols, including x, y, z, R_x, R_y, R_z, and binary products of x, y, z. We will defer discussion of these until later.

The general structure of all character tables is the same as for the C_{2h} table in Figure 2.16. Many other important character tables are listed in Appendix I.

Using the C_{3v} character table given in Figure 2.17, we will illustrate the decomposition of a reducible representation into irreducible representations. This decomposition procedure will be very useful for the applications in later chapters.

C_{3v}	E	$2C_3$	$3\sigma_v$		
A_1	1	1	1	z	$x^2 + y^2, z^2$
A_2	1	1	-1	R_z	
E	2	-1	0	$(x, y)(R_x, R_y)$	$(x^2 - y^2, xy)(xz, yz)$

Figure 2.17
C_{3v} Character Table

In principle it is possible to find a transformation matrix which will block diagonalize a set of representation matrices. However, it can be shown that the irreducible representations in a given reducible representation are given by

$$a_i = \frac{1}{h} \sum_j g_j \chi_j^{IR} \chi_j^{RED}$$

where a_i is the number of times the ith irreducible representation occurs in the reducible representation, and

h is the order of the group under consideration
g_j is the order of the jth class in the group
χ_j^{IR} is the character of the jth class in the irreducible representation
χ_j^{RED} is the character of the jth class in the reducible representation.

Let us choose 5 2 −1 for our reducible representation, $\Gamma_{\text{reducible}}$. There are three terms in the reducible representation because there are three classes of operations in the C_{3v} point group. A reducible representation is merely the sum of the characters of a number of irreducible representations, and a systematic method for determining those irreducible representations follows. We must first multiply term by term the characters of the reducible representation by the characters of each irreducible representation and the order of the class, sum these and then divide by the order of the group. For example, $\Gamma_{\text{reducible}}$ times A_1 times the order of the class divided by the order (h) of the group is shown below:

$$\Gamma_{\text{red.}} \quad 5 \quad 2 \quad -1$$

Number of A_1 irreducible representations equals

$$\frac{1}{6}[(5)(1)(1) + (2)(2)(1) + (-1)(3)(1)] = 1$$

$\Gamma_{\text{red.}}$ times A_2 equals

$$5 \times 1(1) + 2 \times 2(1) + (-1) \times 3(-1) = 12$$
$$12/6 = 2A_2$$

$\Gamma_{\text{red.}}$ times E equals

$$5 \times 1(2) + 2 \times 2(-1) + (-1) \times 3(0) = 10 - 4 + 0 = 6$$
$$6/6 = 1E$$

We now have $\Gamma_{\text{reducible}} = A_1 + 2A_2 + E$ and if we sum the characters of this combination of irreducible representations we will find 5 2 −1 which is equal to the original $\Gamma_{\text{reducible}}$.

In Figure 2.18 the character table for C_{4v} is represented. As our second example of generating irreducible representations from a reducible representation, we will use $\Gamma_{\text{reducible}} = 5\ 1\ 1\ 3\ 1$. There are five terms in the reducible representation because the C_{4v} point group contains five different classes of operations.

$$\Gamma_{\text{red.}} \times A_1$$
$$= 5 \times 1(1) + 1 \times 2(1) + 1 \times 1(1) + 3 \times 2(1) + 1 \times 2(1) = 16$$
$$16/8 = 2A_1$$

$\Gamma_{\text{red.}} \times A_2$

$$= 5 \times 1(1) + 1 \times 2(1) + 1 \times 1(1) + 3 \times 2(-1) + 1 \times 2(-1) = 0$$

$\Gamma_{\text{red.}} \times B_1$

$$= 5 \times 1(1) + 1 \times 2(-1) + 1 \times 1(1) + 3 \times 2(1) + 1 \times 2(-1) = 8$$

$$8/8 = 1B_1$$

$\Gamma_{\text{red.}} \times B_2$

$$= 5 \times 1(1) + 1 \times 2(-1) + 1 \times 1(1) + 3 \times 2(-1) + 1 \times 2(1) = 0$$

$\Gamma_{\text{red.}} \times E$

$$= 5 \times 1(2) + 1 \times 2(0) + 1 \times 1(-2) + 3 \times 2(0) + 1 \times 2(0) = 8$$

$$8/8 = 1E$$

C_{4v}	E	$2C_4$	C_2	$2\sigma_v$	$2\sigma_d$		$h = 8$
A_1	1	1	1	1	1	z	$x^2 + y^2, z^2$
A_2	1	1	1	−1	−1	R_z	
B_1	1	−1	1	1	−1		$x^2 - y^2$
B_2	1	−1	1	−1	1		xy
E	2	0	−2	0	0	(x, y) R_x, R_y	

Figure 2.18
C_{4v} Character Table

When reduced $\Gamma_{\text{reducible}} = 5\ 1\ 1\ 3\ 1$ becomes $2A_1 + B_1 + E$. If we sum these four irreducible representations as a check of our results, we should obtain the original reducible representation. In the following chapters the decomposition of reducible representations and the other principles of the first two chapters will be applied to specific molecular situations.

2-4 SUBGROUPS

Examination of the group multiplication (combination) table for the C_{4v} group, which is displayed in Figure 2.6, will reveal that there are a number of smaller groups within the C_{4v} group. For example, we note that E forms a subgroup by itself; that E, $\overleftarrow{C_4}$, $\overrightarrow{C_4}$ and C_2 form a group of order 4; that E and C_2 form a group of order 2; and so forth. There is a restriction on the order of the subgroups present in a main group, and that is that an integer must result when the order of the subgroup, l, is divided into the order of the main group, h, viz.,

$$\frac{h}{l} = i \quad (i \text{ is an integer})$$

An important consequence of this relationship is that irreducible representations of some group G can be *correlated* with the irreducible representations of some other group G', if G and G' are related by one being a subgroup of the other. There is no problem if the irreducible representations are one-dimensional, for there will be a one to one correlation, but if the irreducible representations are two-dimensional, three-dimensional or larger, there usually are some changes in the dimensions of the irreducible representations from the main group to the subgroup.

Let us examine some examples. We will determine the correlation of the irreducible representations of the subgroup C_{2v} with the irreducible representations of the group C_{4v}. Using the C_{2v} group, we will construct representations by writing down the characters of all the corresponding operations from the C_{4v} group. The representations are

	E	C_2	$\sigma_v(xz)$	$\sigma_v(yz)$	C_{2v} irreducible representations
$\Gamma(A_1)$	1	1	1	1	$= A_1$
$\Gamma(A_2)$	1	1	-1	-1	$= A_2$
$\Gamma(B_1)$	1	1	1	1	$= A_1$
$\Gamma(B_2)$	1	1	-1	-1	$= A_2$
$\Gamma(E)$	2	-2	0	0	$= B_1 + B_2$

We have found the following correlations:

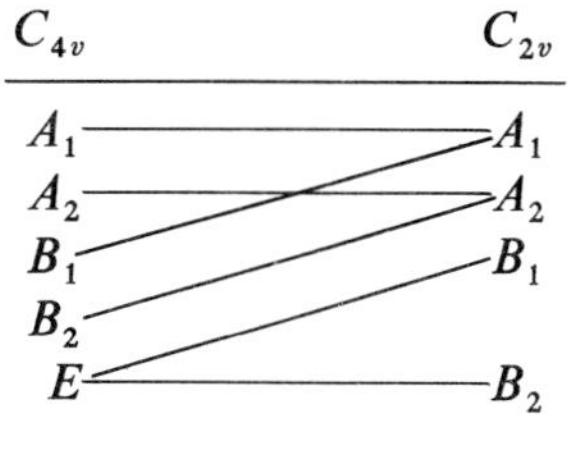

That is, the A_1 and B_1 irreducible representations of the group C_{4v} each correlate with A_1 irreducible representation of C_{2v}, while A_2 and B_2 of C_{4v} each correlate with A_2 of C_{2v}, and E of C_{4v} correlates with $B_1 + B_2$ of C_{2v}.

Consider a second example, that being the C_{3v} and D_{3h} groups. The representations are

	E	$2C_3$	$3\sigma_v$	C_{3v} irreducible representations
$\Gamma(A_1')$	1	1	1	$= A_1$
$\Gamma(A_2')$	1	1	-1	$= A_2$
$\Gamma(E')$	2	-1	0	$= E$
$\Gamma(A_1'')$	1	1	-1	$= A_2$
$\Gamma(A_2'')$	1	1	1	$= A_1$
$\Gamma(E'')$	2	-1	0	$= E$

Although, as we have demonstrated, the determination of the correlation of irreducible representations is a straightforward procedure, it is convenient to have a tabulation of the results for reference. Some important correlation tables are given in Appendix II.

Exercises

2–6: Generate the C_{2v} and C_{4v} character tables.

2–7: Consider a set of unit vectors along the x, y, and z directions in a Cartesian coordinate system. (a) Construct the set of matrices which represent the operations of the C_{2v} group on this set. (b) Are the representation matrices reducible? (c) Construct the representation (reducible or irreducible) spanned by the set of unit vectors $\vec{x}$, $\vec{y}$, $\vec{z}$. (d) If the representation is reducible, decompose it into its irreducible components.

2–8: Repeat Exercise 2–7 for the T_d point group.

2–9: Decompose into irreducible components each of the following reducible representations.

a)

O_h	E	$8C_3$	$6C_2$	$6C_4$	$3C_2$	i	$6S_4$	$8S_6$	$3\sigma_h$	$6\sigma_d$
Γ	5	-1	1	-1	1	5	-1	-1	1	1

b)

T_d	E	$8C_3$	$3C_2$	$6S_4$	$6\sigma_d$
Γ	5	-1	1	-1	1

3

QUANTUM MECHANICAL INTRODUCTION

3-1 BASIC POSTULATES

At this point, it may be helpful to review a few of the basic postulates of wave mechanics. Postulates of any theory are a set of fundamental statements that are believed because they have been justified by experiment; they can not be explained in terms of more fundamental concepts, or the more fundamental concepts become the postulates of the theory. Because these quantum mechanical postulates deal with atomic and molecular properties, they may be difficult to understand in the sense of everyday experience.

The concept of matter waves was first introduced by de Broglie, who stated that a particle, such as an electron, had wavelike properties and that the wave and particle characteristics were related by $\lambda = h/mv$.

In order to simplify notation we will illustrate the mechanics of a particular one-dimensional system, x, rather than a three dimensional one, (x, y, z). Once an understanding of the one-dimensional equations is reached, expanding to three dimensions is a relatively easy transition. According to the first postulate of quantum mechanics a wave function, ψ, belonging to a system containing one particle in one dimension, is a function of x and t. The complete wave function in three-dimensional space would be $\psi(x, y, z, t)$.

The total energy, W, of our one-dimensional system is equal to the sum of the kinetic energy and the potential energy, $W = (1/2m)p_x^2 + V(x)$, where $p_x = mv_x$, the x-component of momentum. From this classical expression for the total energy and according to a second postulate of quantum mechanics, we can obtain the wave equation for the same system by substituting operators for the dynamical variables; a dynamical variable means any property of the system, even though in some cases the property may be a constant. Table 3.1 shows these relationships, where the operators will now operate upon the wave function $\Psi(x, t)$.

Table 3-1
Operators for Some Dynamical Variables

dynamical variable	*operator*	
$f(x)$	$f(x)$	
p_x	$\frac{\hbar}{i}\frac{\partial}{\partial x}$	$\hbar = \frac{h}{2\pi}$
W	$-\frac{\hbar}{i}\frac{\partial}{\partial t}$	

At this point we should discuss briefly the algebra of operators. An operator converts one mathematical function into another. For example, $d/dx(x^3) = 3x^2$ where d/dx operates upon the function x^3. Operators can also act on a function in succession. If we combine d/dx with multiplying by x and operate upon x^3, Equation (3-1) results, where the operations are performed from right to left. However, as shown in Equation (3-2), we do

$$(d/dx)(x)x^3 = (d/dx)x^4 = 4x^3 \tag{3-1}$$

$$(x)(d/dx)x^3 = (x)3x^2 = 3x^3 \tag{3-2}$$

not obtain the same result if we reverse the order of the operations. When the result depends upon the order of the operations, the operators are said to be noncommutative. When two operators X and Y give results that show $XY = YX$, then X and Y are commuting operators.

Linear operators are those for which

$$Op\,[f_1(x) + f_2(x)] = Op\,f_1(x) + Op\,f_2(x).$$

In quantum mechanical discussions all operators are linear.

For a given operator, there may exist a class of functions $f(x)$, which when acted upon by the operator, gives the original function multiplied by some constant k, $Op\,f(x) = k\,f(x)$. An example is shown in Equation (3-3), where $f(x) = e^{kx}$ and $Op = d/dx$. When this is true, $f(x)$ is defined

$$(d/dx)\, e^{kx} = k\, e^{kx} \qquad \textbf{(3-3)}$$

as an eigenfunction of the particular operator Op, and the possible values of k are called the eigenvalues of Op.

Exercises

3–1: Which of the following pairs of operators commute?
- a) $1/x$ and x
- b) $\sqrt{\ }$ and x
- c) cos and x
- d) dx and $1/x$

3–2: Determine whether or not the function is an eigenfunction of the operator. If it is, determine the eigenvalue.

	operator	*function*
a)	d/dx	$8\,e^{5x}$
b)	d^2/dx^2	$A \sin(kx)$
c)	d/dr	$ke^{r/a}$

After the appropriate substitutions the classical equation for the total energy of the system becomes Equation (3-4), which is the Schrödinger wave equation

$$-\frac{\hbar}{i}\frac{\partial \Psi(x,t)}{\partial t} = -\frac{\hbar^2}{2m}\frac{\partial^2 \Psi(x,t)}{\partial x^2} + V(x)\Psi(x,t) \qquad \textbf{(3-4)}$$

for a one-dimensional system. The solutions describe what we believe to be real phenomena; they should, therefore, have all the requirements met by physical waves: $\Psi(x, t)$ and $[\partial\Psi(x, t)/\partial x]$ must be continuous, finite, and singlevalued throughout all space.

3-2 SEPARATION OF VARIABLES

If we look again at Equation (3-4), we find that there are two variables, x and t. These variables can be separated so that $\Psi(x, t) = \psi(x)\phi(t)$ and Equation (3-4) becomes Equation (3-5) where the left side of the equation

$$\frac{1}{\psi(x)}\left(-\frac{\hbar^2}{2m}\frac{\partial^2\psi(x)}{\partial x^2} + V(x)\psi(x)\right) = -\frac{1}{\phi(t)}\frac{\hbar}{i}\frac{\partial\phi(t)}{\partial t} \qquad (3\text{-}5)$$

represents the amplitude of the wave function and the right side the time dependency. If we consider $\psi(x)$ as time independent, then $\psi^*\psi\,dx$ becomes the probability of finding the particle in the interval x to $x + dx$ at a *particular* time. The time independent left side of Equation (3-5) is therefore equal to the total energy of the stationary state as shown in Equation (3-6).

$$\frac{1}{\psi(x)}\left(-\frac{\hbar^2}{2m}\frac{\partial^2\psi(x)}{\partial x^2} + V(x)\psi(x)\right) = W \tag{3-6}$$

Multiplying by $-(2m/\hbar^2)\psi(x)$ and rearranging terms yields Equation (3-7).

$$\frac{\partial^2\psi(x)}{\partial x^2} + \frac{2m}{\hbar^2}(W - V(x))\psi(x) = 0 \tag{3-7}$$

This is Schrödinger's amplitude equation and it will be dealt with more completely in the next chapter. The description of a particle's motion by a wave equation can be thought of only in terms of probability. $\psi^*\psi\,dx$, where ψ^* is the complex conjugate of ψ, is to be interpreted as the probability that a particle is found in the interval x to $x + dx$ at time t. If the wave function ψ is complex, the product of the wave function and its complex conjugate will be positive and real. In cases where the wave function is real, $\psi^*\psi = \psi^2$, and the probability is also positive and real. The probability of finding the particle somewhere in space is 1, which leads us to the next postulate: the integral over all space of $\psi^*\psi$ must equal 1; $\int_{-\infty}^{\infty}\psi^*\psi\,dx = 1$. When this is true, the wave function is said to be normalized.

When the integral $\int_{-\infty}^{\infty}\psi_i^*\psi_j\,dx = 0$, the functions ψ_i and ψ_j are said to be orthogonal. In vector algebra two vectors are orthogonal when they make an angle 90° with one another. When eigenfunctions are orthogonal they are completely independent functions. A set of functions is said to be orthonormal when they are both orthogonal and normalized, such that $\int \psi_i^*\psi_j\,dx = \delta_{ij}$ where δ_{ij} is the Kronecker delta ($\delta_{ij} = 1$ for $i = j$, $\delta_{ij} = 0$ for $i \neq j$).

Exercises

3–3: Normalize the wave function $\psi = A\sin n\pi x/a$. Hint: Use $\int_0^a \psi^2\,dx = 1$ and solve for A.

3–4: Show that $\psi_1 = A\sin \pi x/a$ and $\psi_2 = A\sin 2\pi x/a$ are orthogonal.

The final postulate to be discussed here states that the average value $\bar{\alpha}$, of any dynamical variable α which corresponds to the operator, α_{operator}, is calculated from the wave function by

$$\bar{\alpha} = \int_{-\infty}^{\infty} \psi^* \alpha_{\text{op}} \psi \, dx$$

if ψ is normalized. This equation is most important because it is the one that allows us to calculate observable quantities. The average value, $\bar{\alpha}$, and not a specific value is calculated because of the probability nature of the wave functions.

3-3 BASIS FUNCTIONS

The hydrogen atom problem can be solved exactly, and valid wave functions can be obtained, but for most other problems exact solutions can not be found. To overcome this problem, trial wave functions ϕ, which are approximations of ψ, are used to calculate the energies which correspond to the trial wave functions. A particularly useful sort of trial function is a linear combination of some known set of basis functions, i.e., $\phi = \sum_i c_i \vartheta_i$. A set of basis functions is defined as a set of independent functions which span the space under discussion. The coefficients c_i are varied until the best wave function is obtained. The two most important approximate methods for the evaluation of the coefficients are the variation method and the perturbation method.

3-4 THE PERTURBATION METHOD

The perturbation method utilizes the addition of a term to a Hamiltonian H°, for which exact eigenfunctions ψ° are known, i.e. $H^\circ \psi^\circ = E^\circ \psi^\circ$. The new Hamiltonian would be $H = H^\circ + H'$, where H' provides the perturbation to the original system defined by H°. In practice the computations become

$$H_{ij} = \int \psi_i^{\circ *}(H^\circ + H')\psi_j^\circ \, d\tau = \int \psi_i^{\circ *} H^\circ \psi_j^\circ \, d\tau + \int \psi_i^{\circ *} H' \psi_j^\circ \, d\tau$$

where H_{ij} is the first order corrected energy of the system. The corrected wave function is expressed as

$$\psi_i = \psi_i^\circ + \sum_{i \neq j} c_{ij} \psi_j^\circ$$

where the zeroth-order wave function ψ_i° is improved by the addition of some linear combination of basis functions ψ_j°. Upon substitution into the Hamiltonian equation $H\psi_i = E_i\psi_i$ we get Equation (3-8).

$$(H - E_i)\psi_i^\circ + \sum_{i\neq j} c_{ij}(H - E_i)\psi_j^\circ = 0 \qquad \textbf{(3-8)}$$

The total Hamiltonian H, used in Equation (3-8) can be broken down into H° and the perturbation portion H', which yields Equation (3-9). We

$$(H^\circ + H' - E_i)\psi_i^\circ + \sum_{i\neq j} c_{ij}(H^\circ + H' - E_i)\psi_j^\circ = 0 \qquad \textbf{(3-9)}$$

can eliminate the H° term by using $H^\circ\psi_i^\circ = E_i^\circ\psi_i^\circ$ and

$$(H' + E_i^\circ - E_i)\psi_i^\circ + \sum_{i\neq j} c_{ij}(H' + E_j^\circ - E_i)\psi_j^\circ = 0. \qquad \textbf{(3-10)}$$

If we multiply Equation (3-10) from the left by $\psi_i^{\circ *}$, integrate over all space, and take into account the orthonormality of the ψ°, Equation (3-11) results

$$(H'_{ii} + E_i^\circ - E_i) + \sum_{i\neq j} c_{ij}H'_{ij} = 0 \qquad \textbf{(3-11)}$$

where $H'_{ij} = \int \psi_i^{\circ *}H'\psi_j^\circ\, d\tau$. If we multiply Equation (3-10) from the left by $\psi_k^{\circ *}$, where $k \neq i$, we obtain Equation (3-12). We should note here that the

$$H'_{ki} + c_{ik}(H'_{kk} + E_k^\circ - E_i) + \sum_{i\neq j,k} c_{ij}H'_{kj} = 0 \qquad \textbf{(3-12)}$$

energy and coefficients can be expressed as the sum of the terms of the zeroth, first, second, etc. order perturbations as shown in Equation (3-13)

$$E_i = E_i^\circ + E_i' + E_i'' + \cdots \qquad c_{ij} = c_{ij}' + c_{ij}'' + \cdots \qquad \textbf{(3-13)}$$

It is usually unnecessary to go to higher-order terms.

The first-order perturbation terms may be obtained from Equation (3-11) by neglecting the products of small quantities ($\sum_{i\neq j} c_{ij}H'_{ij}$) giving

$$E_i - E_i^\circ = E_i' = H'_{ii} = \int \psi_i^{\circ *}H'\psi_i^\circ\, dt,$$

which shows that the first-order correction to the energy is expressed in terms of zeroth-order wave functions and the perturbation to the Hamiltonian. Making use of Equation (3-13) the first order terms from Equation (3-12) are

$$H'_{ki} + c_{ik}'(E_k^\circ - E_i^\circ) = 0,$$

from which we can solve for the first-order correction to the coefficients.

$$c'_{ik} = \frac{H'_{ki}}{E_i^\circ - E_k^\circ}$$

The second-order perturbation terms from Equation (3-11) are

$$-E_i'' + \sum_{i \neq j} c'_{ij} H'_{ij} = 0. \qquad \textbf{(3-14)}$$

From Equation (3-14) we can express the second-order correction to the energy as follows

$$E_i'' = \sum_{i \neq j} \frac{H'_{ji} H'_{ij}}{E_i^\circ - E_j^\circ} \qquad \textbf{(3-15)}$$

From Equations (3-14) and (3-15) we can see that the first-order correction to the coefficients and the second-order correction to the energy involve the energy difference. Because these energy terms are in the denominator, the states which have the largest effect on ψ_i° are those which are closest in energy. For a more detailed discussion of degenerate states ($E_i^\circ = E_j^\circ$) and the perturbation method in general, the reader is referred to textbooks dealing with quantum mechanics.

3-5 THE VARIATION METHOD

The basis of the variation method is the principle which states that the energy evaluated from approximate wave functions is never less than the actual energy of the system. The denominator in Equation (3-16) ensures

$$E_\circ \leq E = \frac{\int \phi^* H \phi \, d\tau}{\int \phi^* \phi \, d\tau} \qquad \textbf{(3-16)}$$

that ϕ is normalized. If this equation were not true, we could calculate an energy for the system which would represent a state more stable than the system actually possesses. This principle makes it possible to choose an approximate wave function with variable parameters, c_{ij}, evaluate the energy in terms of these parameters, and then minimize the energy with respect to the parameters by setting the first derivatives, $\partial E/\partial c_{ij}$ equal to zero.

The trial wave function ϕ can be constructed as a linear combination of basis functions

$$\phi = c_1\phi_1 + c_2\phi_2 + \cdots + c_n\phi_n,$$

where the c_i are the variable parameters. If we substitute this linear combination into Equation (3-16) we obtain

$$\sum_{i=1}^{n}\sum_{j=1}^{n} c_i c_j(H_{ij} - ES_{ij}) = 0$$

where $H_{ij} = \int \phi_i^* H\phi_j\, d\tau$ and $S_{ij} = \int \phi_i^*\phi_j\, d\tau$. If the basis functions ϕ are normalized and orthogonal to each other, all $S_{ii} = 1$ and $S_{ij} = 0$ for $i \neq j$. We differentiate with respect to each c and set each $\partial E/\partial c$ equal to zero to obtain Equation (3-17). This can be represented in general by Equation (3-17) or Equation (3-18). This set of simultaneous linear equations that is

$$\frac{\partial E}{\partial c_i} = \sum_{i=1}^{n} c_i(H_{ij} - ES_{ij}) = 0 \tag{3-17}$$

shown in Equation (3-18) forms the *secular equation*. For n equations

$$\begin{aligned} c_1(H_{11} - ES_{11}) + c_2(H_{12} - ES_{12}) + \cdots + c_n(H_{1n} - ES_{1n}) &= 0 \\ c_1(H_{21} - ES_{21}) + c_2(H_{22} - ES_{22}) + \cdots + c_n(H_{2n} - ES_{2n}) &= 0 \\ \vdots \qquad\qquad\qquad\qquad\qquad\qquad & \vdots \\ c_1(H_{n1} - ES_{n1}) + c_2(H_{n2} - ES_{n2}) + \cdots + c_n(H_{nn} - ES_{nn}) &= 0 \end{aligned} \tag{3-18}$$

all equal to zero in n unknown variables, the only solution (except for the trivial solution where all $c_i = 0$) is obtained when the determinant of the coefficients equals zero. This determinantal equation is called the *secular equation* and is shown in Equation (3-19). The secular equation can be solved to obtain the energies E, and each energy can be substituted into Equation (3-18) to solve for the coefficients c_i which provide the best trial function that can be formed from the chosen basis set.

$$\begin{vmatrix} H_{11} - ES_{11} & H_{12} - ES_{12} & \cdots & H_{1n} - ES_{1n} \\ H_{21} - ES_{21} & H_{22} - ES_{22} & \cdots & H_{2n} - ES_{2n} \\ \vdots & & & \vdots \\ H_{n1} - ES_{n1} & H_{n2} - ES_{n2} & \cdots & H_{nn} - ES_{nn} \end{vmatrix} = 0 \tag{3-19}$$

If the discussion deals with the formation of molecules from atoms, then the basis functions (atomic orbitals) ϕ are orthogonal to each other if they are centered on the same atom, and $S_{ij} = 0$. However, if the atomic orbitals are on different atoms, S_{ij} becomes the overlap integral which may or may not be zero.

Using the variation method to solve for the energies for a two-term wave function should be helpful. We will first write $\phi = c_1\phi_A + c_2\phi_B$ where ϕ_A and ϕ_B are the basis functions from which ϕ is derived. By applying the variation principle and using Equation (3-18) we generate the equations shown in Equation (3-20). We can simplify these equations by

$$\begin{aligned} c_1(H_{AA} - ES_{AA}) + c_2(H_{AB} - ES_{AB}) &= 0 \\ c_2(H_{BA} - ES_{BA}) + c_2(H_{BB} - ES_{BB}) &= 0 \end{aligned} \qquad \textbf{(3-20)}$$

realizing that $H_{AB} = H_{BA}$ and remembering that $S_{AA} = S_{BB} = 1$. Also, to make the calculation some what more simple, if ϕ_A and ϕ_B are equivalent basis functions, $H_{AA} = H_{BB}$. The secular equation for the simplified equations is shown in Equation (3-21). This equation can be written

$$\begin{vmatrix} H_{AA} - E & H_{AB} - ES_{AB} \\ H_{AB} - ES_{AB} & H_{AA} - E \end{vmatrix} = 0 \qquad \textbf{(3-21)}$$

as $(H_{AA} - E)^2 = (H_{AB} - ES_{AB})^2$ or $H_{AA} - E = \pm(H_{AB} - ES_{AB})$. The two values for E which result are $E = (H_{AA} + H_{AB})/(1 + S_{AB})$ and $E = (H_{AA} - H_{AB})/(1 - S_{AB})$.

If we substitute $E = (H_{AA} + H_{AB})/(1 + S_{AB})$ into Equations (3-20), the result is what is shown in Equation (3-22). If we simplify these equations

$$\begin{aligned} c_1\left(H_{AA} - \frac{H_{AA} + H_{AB}}{1 + S_{AB}}\right) + c_2\left(H_{AB} - S_{AB}\frac{H_{AA} + H_{AB}}{1 + S_{AB}}\right) &= 0 \\ c_1\left(H_{AB} - S_{AB}\frac{H_{AA} + H_{AB}}{1 + S_{AB}}\right) + c_2\left(H_{AA} - \frac{H_{AA} + H_{AB}}{1 + S_{AB}}\right) &= 0 \end{aligned} \qquad \textbf{(3-22)}$$

we obtain the following:

$$\begin{aligned} &c_1(H_{AA} + H_{AA}S_{AB} - H_{AA} - H_{AB}) \\ &\qquad + c_2(H_{AB} + H_{AB}S_{AB} - S_{AB}H_{AA} - S_{AB}H_{AB}) = 0 \\ &c_1(H_{AB} + H_{AB}S_{AB} - S_{AB}H_{AA} - S_{AB}H_{AB}) \\ &\qquad + c_2(H_{AA} + H_{AA}S_{AB} - H_{AA} - H_{AB}) = 0 \end{aligned}$$

or

$$c_1(H_{AA}S_{AB} - H_{AB}) + c_2(H_{AB} - S_{AB}H_{AA}) = 0$$
$$c_1(H_{AB} - S_{AB}H_{AA}) + c_2(H_{AA}S_{AB} - H_{AB}) = 0$$

Both equations yield the solution $c_1 = c_2$. Therefore, $\phi = c_1\phi_A + c_2\phi_B$ and when ϕ is normalized, in the zero overlap approximation $c_1 = c_2 = 1/\sqrt{2}$.

The elementary concepts of quantum mechanics that are introduced in this chapter and the use of the variation method for approximating energy values will be applied to specific examples in the following chapters.

Bibliography

Of the books on quantum mechanics, it is suggested that the following may be appropriate:

C. A. Coulson, *Valence*, 2d ed. New York: Oxford University Press, 1961.

H. Eyring, J. Walter, and G. E. Kimball, *Quantum Chemistry*, New York: John Wiley and Sons, 1944.

L. Pauling and E. B. Wilson, *Introduction to Quantum Mechanics*, New York: McGraw-Hill Book Co., 1935.

4

ATOMIC AND HYBRID ORBITALS

The techniques of symmetry which have been developed in the earlier chapters can be effectively used to determine the bonding orbitals in a variety of molecular geometries. However, before we discuss hybrid orbitals we shall describe some of the theory pertaining to atomic orbitals and the symmetries of these orbitals. The symmetries of several atomic orbitals will then be determined, and a procedure for constructing hybrid orbitals from combinations of atomic orbitals will be developed.

4-1 ATOMIC ORBITALS

As an introduction to atomic orbitals we shall describe the solution of the Schrödinger Equation (4-1) for the simplest atom, H, and then point out the approximations which are required for more complicated, many-electron, atoms. Even here we have made an approximation, since Equation (4-1) describes the electron with respect to a fixed nucleus.

$$\nabla^2\psi + \frac{2m}{\hbar^2}(W - V)\psi = 0 \qquad \textbf{(4-1)}$$

where W is the total energy

V is the potential energy

$\hbar = h/2\pi$

ψ is a wave function which describes the electron

∇^2 is $\frac{\partial^2}{\partial x^2} + \frac{\partial^2}{\partial y^2} + \frac{\partial^2}{\partial z^2}$

$-\frac{\hbar^2}{2m}\nabla^2$ is the kinetic energy operator and m is the electron mass.

For this problem it is convenient to use polar coordinates. The necessary relationships between cartesian and polar coordinates are shown in Figure 4.1. In the hydrogen atom the single electron is subject to the coulombic

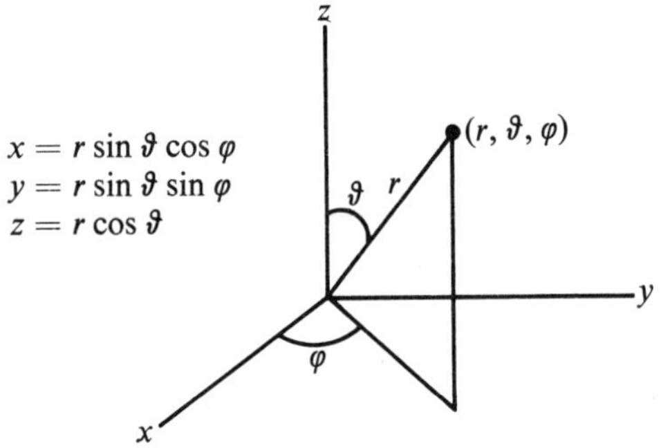

Figure 4.1

Relationship of the Cartesian Coordinates to Polar Coordinates

attraction (Coulomb field) of the positive nucleus. The potential energy of this system of one electron and one nucleus is given by $V = -e^2/r$. The wave function which describes this electron is written as $\Psi(r, \theta, \phi) = R(r)\Theta(\theta)\Phi(\phi)$, where $R(r)$ denotes the radial dependence, and the factor $\Theta(\theta)\Phi(\phi)$ denotes the angular dependence of the electronic distribution. Although we will not go into the details here, it is a very instructive exercise for the student to transform ∇^2 from the cartesian coordinates given in (4-1) to polar coordinates and to show that

$$\nabla^2 = \frac{1}{r^2}\frac{\partial}{\partial r}\left(r^2\frac{\partial}{\partial r}\right) + \frac{1}{r^2\sin^2\theta}\frac{\partial^2}{\partial\phi^2} + \frac{1}{r^2\sin\theta}\frac{\partial}{\partial\theta}\sin\theta\frac{\partial}{\partial\theta}$$

Thus, the Schrödinger equation becomes

$$\frac{1}{r^2}\frac{\partial}{\partial r}\left(r^2\frac{\partial}{\partial r}R\Theta\Phi\right)+\frac{1}{r^2\sin^2\theta}\frac{\partial^2}{\partial\phi^2}R\Theta\Phi+\frac{1}{r^2\sin\theta}\frac{\partial}{\partial\theta}\sin\theta\frac{\partial}{\partial\theta}R\Theta\Phi$$
$$+\frac{2m}{\hbar^2}[W-V]\,R\Theta\Phi=0$$

In order to handle this equation, the variables $r\theta\phi$ are separated in the following manner: First the equation is divided by $R\Theta\Phi$ to give

$$\frac{1}{r^2R}\frac{\partial}{\partial r}\left(r^2\frac{\partial R}{\partial r}\right)+\frac{1}{\Phi r^2\sin^2\theta}\frac{\partial^2\Phi}{\partial\phi^2}+\frac{1}{\Theta r^2\sin\theta}\frac{\partial}{\partial\theta}\left(\sin\theta\frac{\partial\Theta}{\partial\theta}\right)$$
$$+\frac{2m}{\hbar^2}[W-V]=0$$

The next step is multiplication by $r^2\sin^2\theta$, and this yields

$$\frac{\sin^2\theta}{R}\frac{\partial}{\partial r}\left(r^2\frac{\partial R}{\partial r}\right)+\frac{1}{\Phi}\frac{\partial^2\Phi}{\partial\phi^2}+\frac{\sin\theta}{\Theta}\frac{\partial}{\partial\theta}\left(\sin\theta\frac{\partial\Theta}{\partial\theta}\right)$$
$$+\frac{2mr^2\sin^2\theta}{\hbar^2}[W-V]=0$$

and upon transposing the term involving Φ we get

$$\frac{\sin^2\theta}{R}\frac{\partial}{\partial r}\left(r^2\frac{\partial R}{\partial r}\right)+\frac{\sin\theta}{\Theta}\frac{\partial}{\partial\theta}\left(\sin\theta\frac{\partial\Theta}{\partial\theta}\right)+\frac{2m}{\hbar^2}r^2\sin^2\theta[W-V]=-\frac{1}{\Phi}\frac{\partial^2\Phi}{\partial\phi^2}$$

In this equation the variables r and θ are only on the left side and ϕ is only on the right. As a consequence, no matter what values r, θ, and ϕ take independently, the sum of the terms on the left must equal the term on the right. This can only be so if each side of the equation is equal to the same constant, say m^2. We can immediately write Equation (4-2a) in terms of the variable ϕ.

$$\frac{1}{\Phi}\frac{d^2\Phi}{d\phi^2}=-m^2 \tag{4-2a}$$

The left side of Equation (4-1), which is also equal to the constant m^2, yields

$$\frac{\sin^2\theta}{R}\frac{\partial}{\partial r}\left(r^2\frac{\partial R}{\partial r}\right)+\frac{\sin\theta}{\Theta}\frac{\partial}{\partial\theta}\left(\sin\theta\frac{\partial\Theta}{\partial\theta}\right)+\frac{2mr^2\sin^2\theta}{\hbar^2}[W-V]=m^2 \tag{4-2b}$$

Upon dividing through by $\sin^2\theta$ and rearranging we get

$$\frac{1}{R}\frac{\partial}{\partial r}\left(r^2\frac{\partial R}{\partial r}\right) + \frac{2mr^2}{\hbar^2}[W - V] = \frac{m^2}{\sin^2\theta} - \frac{1}{\Theta \sin\theta}\frac{\partial}{\partial\theta}\left(\sin\theta\frac{\partial\Theta}{\partial\theta}\right) \qquad \textbf{(4-2c)}$$

Now each side of the equation contains only one independent variable, and, as above, each side must be equal to the same constant, say β.

Upon multiplying the equation obtained by setting the right side of Equation (4-2c) equal to β by θ and transposing, we obtain

$$\frac{m^2\Theta}{\sin^2\theta} - \frac{1}{\sin\theta}\frac{d}{d\theta}\left(\sin\theta\frac{d\Theta}{d\theta}\right) - \beta\Theta = 0 \qquad \textbf{(4-3)}$$

and upon multiplying the equation obtained by setting the left side of Equation (4-2c) equal to β by R, dividing by r^2 and transposing, we get

$$\frac{1}{r^2}\frac{d}{dr}\left(r^2\frac{dR}{dr}\right) - \frac{\beta}{r^2}R + \frac{2m}{\hbar^2}[W - V]R = 0 \qquad \textbf{(4-4)}$$

We are now in a position to consider the solution of the differential Equations (4-2a), (4-3), and (4-4). First consider the Φ Equation (4-2a). A possible solution is $\Phi_m(\phi) = Ce^{\pm im\phi}$. To demonstrate that this is a solution we must show that

$$\frac{d^2\Phi}{d\phi^2} = -m^2\Phi$$

The first derivative is

$$\frac{d}{d\phi}Ce^{\pm im\phi} = \pm imCe^{\pm im\phi}$$

and

$$\frac{d^2\Phi}{d\phi^2} = \frac{d}{d\phi}(\pm imCe^{\pm im\phi}) = (\pm im)(\pm im)Ce^{\pm im\phi} = -m^2Ce^{\pm im\phi} = -m^2\Phi$$

The constant C, which is the normalization constant, may now be evaluated. The limits on the variable ϕ are zero and 2π, and by definition of normalization

$$\int_0^{2\pi} \Phi^*\Phi\, d\phi = 1$$

This gives

$$\int_0^{2\pi} C^2 e^{\mp im\phi} e^{\pm im\phi}\, d\phi = 1$$

$$\int_0^{2\pi} C^2\, d\phi = 1$$

$$2\pi C^2 = 1$$

$$C = \frac{1}{\sqrt{2\pi}}$$

so that $\Phi_m(\phi) = (1/\sqrt{2\pi})e^{\pm im\phi}$ where $m = 0, \pm 1, \pm 2, \ldots$. It is necessary for $|m|$ to be an integer for the function to be single valued at the point $\phi = 0$ which, of course, is equivalent to the point $\phi = 2\pi$.

The solution of the θ Equation (4-3) is not as simple as the solution of the Φ equation, but it can be shown that the normalized equation is

$$\Theta_l^m(\theta) = \sqrt{\frac{(2l+1)(l-|m|)!}{2(l+|m|)!}}\mathrm{P}_l^{|m|}(\cos\theta)$$

where $l(l+1) = \beta$ and $\mathrm{P}_l^{|m|}$ is the associated Legendre function. Also, we see that $|m|$ can not be greater than the quantum number l, which must be an integer, or we would have the factorial of a negative number. Factorials of negative numbers are undefined.

It is instructive to consider the solution of the radial Equation (4-4). For large values of r, the radial equation reduces to

$$\frac{d^2R}{dr^2} + \frac{2m}{\hbar^2}WR = 0$$

Since for all hydrogen-like ions $V = -Ze^2/r$, where Z is the nuclear charge, we may rewrite this as

$$\frac{d^2R}{dr^2} = -\frac{2m}{\hbar^2}WR$$

Letting $2mW/\hbar^2 = -\alpha^2$, $R = e^{\pm\alpha r}$ becomes a solution since

$$\frac{dR}{dr} = \pm\alpha e^{\pm\alpha r} \qquad \text{and} \qquad \frac{d^2R}{dr^2} = \alpha^2 e^{\pm\alpha r}$$

The solution with the positive exponent must be discarded since the wave function would be ill-behaved, *i.e.* no longer be finite, as the variable r approaches infinity.

At this point we should check to see if the negative solution satisfies the complete radial equation for the case $l = 0$. (In practice we are setting $\beta = 0$.) The equation is

$$\frac{1}{r^2}\frac{d}{dr}\left(r^2\frac{d(e^{-\alpha r})}{dr}\right) + \frac{2m}{\hbar^2}\left(W + \frac{Ze^2}{r}\right)e^{-\alpha r} = 0$$

and by carrying out the indicated operations

$$\frac{1}{r^2}\frac{d}{dr}r^2(-\alpha)e^{-\alpha r} + \frac{2m}{\hbar^2}\left(W + \frac{Ze^2}{r}\right)e^{-\alpha r} = 0$$

$$-\frac{\alpha}{r^2}\{2re^{-\alpha r} - \alpha r^2 e^{-\alpha r}\} + \ldots = 0$$

$$-\frac{2\alpha e^{-\alpha r}}{r} + \alpha^2 e^{-\alpha r} + \frac{2mW}{\hbar^2}e^{-\alpha r} + \frac{2mZe^2}{\hbar^2 r}e^{-\alpha r} = 0$$

Substituting back in for α^2 in the second term, there results

$$-\frac{2\alpha e^{-\alpha r}}{r} + \frac{2mZe^2}{\hbar^2 r}e^{-\alpha r} = 0$$

In order for this to be a solution α must be defined as

$$\alpha = \frac{mZe^2}{\hbar^2} \qquad \text{or} \qquad \alpha^2 = \frac{m^2Z^2e^4}{\hbar^4}$$

Substituting this in for α^2 in our original equation, we get

$$-\frac{2mW}{\hbar^2} = \frac{m^2Z^2e^4}{\hbar^4} \qquad \text{or} \qquad W = -\frac{mZ^2e^4}{2\hbar^2}$$

This is the energy expression which results from the Bohr model of the hydrogen atom.

To normalize the radial equation we let $a_o = \hbar^2/me^2$, remembering that $\alpha = (mZe^2)/\hbar^2$ and write

$$\int_0^\infty Ne^{-\alpha r} \cdot Ne^{-\alpha r} r^2 dr = 1$$

$$= N^2 \int_0^\infty e^{-2Zr/a_0} r^2 dr = 1$$

From a mathematical handbook of integrals we find

$$\int_0^\infty x^n e^{-ax}\,dx = \frac{\Gamma(n+1)}{a^{n+1}}$$

Here $a = 2Z/a_o$ and $n = 2$. $\Gamma(n+1) = n\Gamma(n)$ and $\Gamma(n) = (n-1)!$ Therefore $\Gamma(2+1) = 2\cdot 1 = 2$ and $N^2\{2/(2Z/a_o)^3\} = 1$. This gives $N = 2\sqrt{(Z/a_o)^3}$.

The general form for the solution of the radial equation is

$$R_n^l(r) = \sqrt{\left(\frac{2Z}{na_o}\right)^3 \frac{(n-l-1)!}{2n[(n+l)!]^3}}\, e^{-\rho/2}\rho^l L_{n+l}^{2l+1}(\rho)$$

where $\rho = (2Z/na_o)r$ and $L_{n+l}^{2l+1}(\rho)$ represents the associated Laguerre polynomials. Note that the term $(n-l-1)!$ requires that the maximum value of l is $n-1$ where n is required to be an integer. Larger values of l would give an undefined negative factorial. The quantum numbers nlm which have resulted in this solution are called principal, azimuthal, and magnetic quantum numbers, respectively.

Using the notation Y_l^m for the angular part of the electronic wave functions, let us now write down a few angular wave functions and examine their properties.

$$\begin{aligned}
Y_0^0 &= \frac{1}{\sqrt{2\pi}} && s \text{ orbital}\\
Y_1^0 &= \frac{1}{2}\sqrt{\frac{3}{\pi}}\cos\theta && \\
& && p \text{ orbitals}\\
Y_1^{\pm 1} &= \frac{1}{2}\sqrt{\frac{3}{2\pi}}\sin\theta\, e^{\pm i\phi} && \\
Y_2^0 &= \frac{1}{4}\sqrt{\frac{5}{\pi}}(3\cos^2\theta - 1) && \\
Y_2^{\pm 1} &= \frac{1}{4}\sqrt{\frac{15}{2\pi}}\sin 2\theta\, e^{\pm i\phi} && d \text{ orbitals}\\
Y_2^{\pm 2} &= \frac{1}{4}\sqrt{\frac{15}{2\pi}}\sin^2\theta\, e^{\pm 2i\phi} &&
\end{aligned} \qquad (4\text{-}5)$$

Since the $Y_{1,2}^{\pm 1}$ and $Y_2^{\pm 2}$ functions have imaginary exponents, they can not be illustrated in real space. In order to illustrate these in real space, real linear combinations of angular functions with the same n and l are constructed. For example, for one of the p orbitals the linear combination is

$$\sqrt{\frac{1}{2}}(Y_1^1 + Y_1^{-1}) = \sqrt{\frac{1}{2}}\left(\frac{1}{2}\sqrt{\frac{3}{2\pi}}\sin\theta\, e^{i\phi} + \frac{1}{2}\sqrt{\frac{3}{2\pi}}\sin\theta\, e^{-i\phi}\right)$$
$$= \frac{1}{2}\sqrt{\frac{3}{\pi}}\sin\theta\left(\frac{e^{i\phi} + e^{-i\phi}}{2}\right)$$
$$= \frac{1}{2}\sqrt{\frac{3}{\pi}}\sin\theta\cos\phi$$

since

$$\cos\phi = (e^{i\phi} + e^{-i\phi})/2$$

This function can be illustrated in real space. It can be seen that this is the x-component of r, and therefore, the orbital which this angular function describes is called p_x. The wave function for a specific p_x orbital will be the product of this angular function and the appropriate radial function. As an exercise the reader should work out the expressions for the angular parts of the p_y orbital and of the various d-orbitals which are listed in Table 4.1.

Table 4.1
Angular Wave Functions

Orbital	*Function*
s	$1/2\pi$
p_x	$[(\sqrt{3/\pi})/2]\sin\vartheta\cos\varphi$
p_y	$[(\sqrt{3/\pi})/2]\sin\vartheta\sin\varphi$
p_z	$[(\sqrt{3/\pi})/2]\cos\vartheta$
d_z^2	$[(\sqrt{5/\pi})/4](3\cos^2\vartheta - 1)$
d_{xz}	$[(\sqrt{15/\pi})/2]\sin\vartheta\cos\vartheta\cos\varphi$
d_{yz}	$[(\sqrt{15/\pi})/2]\sin\vartheta\cos\vartheta\sin\varphi$
$d_{x^2-y^2}$	$[(\sqrt{15/\pi})/4]\sin^2\vartheta\cos 2\varphi$
d_{xy}	$[(\sqrt{15/\pi})/4]\sin^2\vartheta\sin 2\varphi$

4-2 SYMMETRIES OF ATOMIC ORBITALS

In the previous section it was shown that a linear combination of Y_1^{+1} and Y_1^{-1} resulted in a function which corresponded to the x-component of r and therefore was called p_x. The real forms of some common angular wave functions are listed in Table 4.1. Inspection of the table will show that the forms for the orbital functions identified as p_y and p_z are the y and z components, respectively, of r and the designation of these orbitals as p_y and p_z becomes clear.

One more example should suffice to clarify the procedure. Consider the orbital which has the angular function $\sin^2 \theta \sin 2\phi$. Since $\sin 2\phi = 2 \sin \phi \cos \phi$, we have $2(\sin\theta \sin\phi)(\sin\theta\cos\phi)$ or $2(y/r)(x/r) = (2/r)xy$. Thus, the designation d_{xy} for the orbital with the real angular form $\sin^2\theta \sin 2\phi$ may now be understood. The function for a specific d_{xy} orbital will be the product of this angular function and the appropriate radial function.

Let us now consider the effect of the symmetry operations of the point group C_{3v} on the set of functions p_x, p_y, p_z. Note that, as can be seen from Figure 4.2, no operation in the point group C_{3v} changes the value of θ, but that ϕ is changed by C_3^1 and σ_v in the following ways:

$$C_3^1 \phi_1 = \phi_2 = \phi_1 + 2\pi/3$$
$$\sigma_v \phi_1 = \phi_2 = -\phi_1$$

Therefore,

$$C_3^1 p_x = C_3^1 (R \sin\theta \cos\phi) = R \sin\theta \cos(\phi + 2\pi/3)$$

and

$$C_3^1 p_y = C_3^1 (R \sin\theta \sin\phi) = R \sin\theta \sin(\phi + 2\pi/3)$$

From the trigonometric relationships

$$\cos(\alpha \pm \beta) = \cos\alpha \cos\beta \mp \sin\alpha \sin\beta$$
$$\sin(\alpha \pm \beta) = \sin\alpha \cos\beta \pm \cos\alpha \sin\beta$$

Therefore

$$\begin{aligned} C_3^1(p_x) &= R \sin\theta (\cos\phi \cos 2\pi/3 - \sin\phi \sin 2\pi/3) \\ &= R \sin\theta (-\tfrac{1}{2}\cos\phi - [\sqrt{3}/2] \sin\phi) \\ &= -\tfrac{1}{2} p_x - (\sqrt{3}/2) p_y \end{aligned}$$

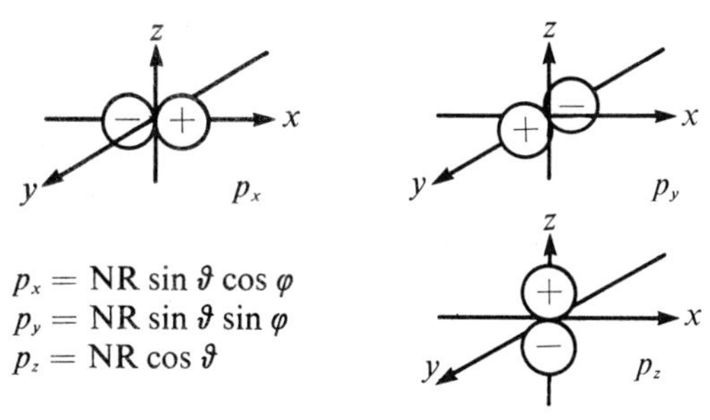

Figure 4.2
Angular Wave Functions for p Orbitals

while

$$\begin{aligned} C_3^1(p_y) &= R \sin\theta\,(\sin\phi \cos 2\pi/3 + \cos\phi \sin 2\pi/3) \\ &= R\sin\theta\,(-\tfrac{1}{2}\sin\phi + [\sqrt{3}/2]\cos\phi) \\ &= -\tfrac{1}{2}p_y + (\sqrt{3}/2)\,p_x \end{aligned}$$

and

$$C_3^1(p_z) = p_z$$

The representation matrix for the C_3 operation is

$$C_3^1\begin{pmatrix} p_x \\ p_y \\ p_z \end{pmatrix} = \begin{pmatrix} -\frac{1}{2} & -\sqrt{3}/2 & 0 \\ +\sqrt{3}/2 & -\frac{1}{2} & 0 \\ 0 & 0 & 1 \end{pmatrix}\begin{pmatrix} p_x \\ p_y \\ p_z \end{pmatrix}$$

For the reflection σ_v^{xz} the representation matrix is

$$\sigma_v^{xz}\begin{pmatrix} p_x \\ p_y \\ p_z \end{pmatrix} = \begin{pmatrix} 1 & 0 & 0 \\ 0 & -1 & 0 \\ 0 & 0 & 1 \end{pmatrix}\begin{pmatrix} p_x \\ p_y \\ p_z \end{pmatrix}$$

Collecting the characters of the representation matrices yields

C_{3v}	E	$2C_3$	$3\sigma_v$	
$\Gamma(p_x, p_y p_z)$	3	0	1	$= A_1 + E$

Comparison of these results with the representation spanned by the unit vectors $\vec{x}$, $\vec{y}$, $\vec{z}$ (see Exercise 2–7) will reveal that the set of functions p_x, p_y, p_z transform like x, y, z. This is a general result. We can determine how the p-orbitals transform in any point group by consulting the character table for that group and seeing how x, y, z transform. As an example, in C_{2v} x transforms as B_1, y as B_2, and z as A_1; therefore, p_x transforms as B_1, p_y as B_2, and p_z as A_1. The same reasoning is extended to the d-orbitals; that is, d_{xy} transforms as the binary product xy while d_z^2 goes as $3z^2 - r^2$. The s-orbital always transforms as the totally symmetric representation of the point group since the orbital itself is spherically symmetrical.

Exercises

4–1: Before attempting to complete this exercise read:
B. Perlmutter-Hayman, *J. Chem. Ed.*, *46*, 428 (1969);

E. A. Ogryzlo and G. B. Porter, *ibid.*, *40*, 256 (1963); A. C. Wahl, *Science*, *151*, 961 (1966).

a) The angular wave function for a hydrogenic p_z orbital is $[(\sqrt{3/\pi})/2] \cos \phi$. Plot this function.

b) The hydrogenic radial wave function for $n = 2$, $l = 1$ is

$$R_{21}(r) = \frac{(Z/a_o)^{3/2}}{2\sqrt{6}} \rho \exp(-\rho/2)$$

where $\rho = 2Zr/na_o$ and a_o is the Bohr radius. Plot this function for a hydrogen atom.

c) According to the postulates of quantum mechanics the probability of finding the electron at any point is given by $\psi^*\psi$. Explain how the information obtained from the plots in a) and b) and this postulate can be used to describe the spatial distribution of an electron in the $2p_z$ orbital of the hydrogen atom.

4–2: How does the *d*-orbital $x^2 - y^2$ transform in the point groups C_{2v}, D_{2d}, O_h, T_d, D_{3h}?

4–3: a) The angular parts, except for numerical factors, for the wave functions with $l = 3$ are

$$\psi_0: \quad (5 \cos^3 \theta - 3 \cos \theta) \qquad m_l = 0$$

$$\psi_{\pm 1}: \quad \sin \theta(5 \cos^2 \theta - 1)e^{\pm i\phi} \qquad m_l = \pm 1$$

$$\psi_{\pm 2}: \quad (\sin^2 \theta \cos \theta)e^{\pm 2i\phi} \qquad m_l = \pm 2$$

$$\psi_{\pm 3}: \quad (\sin^3 \theta)e^{\pm 3i\phi} \qquad m_l = \pm 3$$

Generate real forms of these functions.

b) The angular wave functions in Cartesian coordinates and in polar coordinates are given below for the *f*-orbitals. Demonstrate that the indicated identities hold true.

$$z(5z^2 - 3r^2) \equiv 5 \cos^3 \theta - 3 \cos \theta$$

$$x(5z^2 - r^2) \equiv \sin \theta(5 \cos^2 \theta - 1) \cos \phi$$

$$y(5z^2 - r^2) \equiv \sin \theta(5 \cos^2 \theta - 1) \sin \phi$$

$$z(xy) \equiv \sin^2 \theta \cos \theta \sin 2\phi$$

$$z(x^2 - y^2) \equiv \sin^2 \theta \cos \theta \cos 2\phi$$

$$x(x^2 - 3y^2) \equiv \sin^3 \theta \cos 3\phi$$

$$y(3x^2 - y^2) \equiv \sin^3 \theta \sin 3\phi$$

c) For which irreducible representations of the point groups T_d, O_h, D_{4h} do the seven functions form bases?
d) Determine the shapes of the *f*-orbitals.

4-3 SIGMA HYBRID ORBITALS IN TETRAHEDRAL MOLECULES

The procedure used for determining the atomic orbitals which participate in the formation of hybrid orbitals is straightforward. First, the reducible representation which is spanned by the set of equivalent chemical bonds is constructed and reduced to its irreducible components. Atomic orbitals which transform according to the various irreducible representations are then selected to be combined to form the set of hybrid orbitals. The procedure becomes clear by consideration of an example.

Consider the tetrahedral molecule CCl_4 shown in Figure 4.3. This molecule is representative of a number of tetrahedral molecules which belong to the point group T_d, and the hybrid orbitals which will be constructed below for CCl_4 will be correct for any tetrahedral molecule. It can be noted that the symmetry of an atomic orbital is independent of the principal quantum number, n, so it is expected that the orbitals to be selected will depend on the atomic number of the central atom. For example, silicon in $SiCl_4$ will use 3*s* and 3*p* orbitals while carbon in CCl_4 will use 2*s* and 2*p* orbitals.

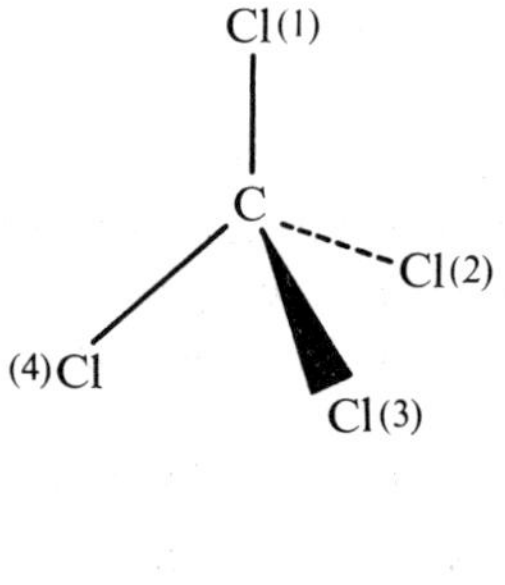

Figure 4.3
Tetrahedral CCl_4

The first step is to construct the set of representation matrices for the group symmetry operations using the set of four carbon-chlorine bonds as a basis set. In T_d the classes are E, 8 C_3, 3 C_2, 6 S_4, and 6 σ_d. Since we are interested only in the reducible representation it is necessary only to construct a representation matrix for one member in each class. Recall that the character, χ, (sum of the elements on the diagonal of the matrix), of representation matrices which belong to the same class are equal. The representation matrices are shown in Figure 4.4. The reducible representation is

T_d	E	$8C_3$	$3C_2$	$6S_4$	$6\sigma_d$
$\Gamma_{\text{C-Cl}}$	4	1	0	0	2

which, according to the procedures worked out in Section 2-3, yields $A_1 + T_2$. Consultation of the T_d character table reveals that the s orbital transforms as A_1 and the sets (x, y, z) and (xy, xz, yz) transform as T_2.

$$E\begin{pmatrix}1\\2\\3\\4\end{pmatrix}=\begin{pmatrix}1&0&0&0\\0&1&0&0\\0&0&1&0\\0&0&0&1\end{pmatrix}\begin{pmatrix}1\\2\\3\\4\end{pmatrix}\qquad C_3\begin{pmatrix}1\\2\\3\\4\end{pmatrix}=\begin{pmatrix}1&0&0&0\\0&0&1&0\\0&0&0&1\\0&1&0&0\end{pmatrix}\begin{pmatrix}1\\2\\3\\4\end{pmatrix}$$

$$C_2\begin{pmatrix}1\\2\\3\\4\end{pmatrix}=\begin{pmatrix}0&1&0&0\\1&0&0&0\\0&0&0&1\\0&0&1&0\end{pmatrix}\begin{pmatrix}1\\2\\3\\4\end{pmatrix}\qquad S_4\begin{pmatrix}1\\2\\3\\4\end{pmatrix}=\begin{pmatrix}0&0&1&0\\0&0&0&1\\1&0&0&0\\0&1&0&0\end{pmatrix}\begin{pmatrix}1\\2\\3\\4\end{pmatrix}$$

$$\sigma_d\begin{pmatrix}1\\2\\3\\4\end{pmatrix}=\begin{pmatrix}1&0&0&0\\0&1&0&0\\0&0&0&1\\0&0&1&0\end{pmatrix}\begin{pmatrix}1\\2\\3\\4\end{pmatrix}$$

Figure 4.4
T_d Representation Matrices

There is a short cut which can be used to construct the reducible representation spanned by the set of sigma hybrid orbitals. The character of each operation is given by the sum of the bonds which are not moved by the operation, for they are the only ones that remain on the diagonal of the representation matrix and thus contribute to the character. The characters are

$E = 4$ since no bonds are shifted

$C_3 = 1$ since three bonds are shifted

$C_2 = 0$ since all bonds are shifted

$S_4 = 0$ since all bonds are shifted

$\sigma_d = 2$ since two bonds are shifted

For a molecule like CCl_4, the choice of carbon orbitals to be used in bonding would be sp^3, where the symbolism says that one s orbital and three p orbitals are combined to give four sp^3 hybrids. It is important to realize that from the standpoint of symmetry the set sd^3 is equally as correct, or that a combination of sd^3 and sp^3 orbitals will fulfill the symmetry requirements. However, from energetic arguments, it is not expected that $3d$ orbitals will be used by carbon atoms since the $3d$ orbitals are so high in energy relative to $2p$. The situation is not nearly as clear-cut for $SiCl_4$. Perhaps the d orbitals are extensively used in bonding since they differ little in energy from $3p$ orbitals.

The combinations of central atom orbitals which are used in the construction of hybrid orbitals have been tabulated in many places. As an exercise the reader should show that the following combinations may occur for sigma bonding in commonly occurring geometries:

trigonal planar molecule	sp^2, sd^2, dp^2, d^3
square planar molecule	dsp^2, d^2p^2
octahedral molecule	d^2sp^3

4-4 ANALYTICAL FORMS FOR HYBRID ORBITALS

In the previous section the combinations of atomic orbitals used in the construction of hybrid orbitals were determined by group theoretical methods. In this section a procedure for the construction of the analytical forms of the hybrid orbitals will be described. For the demonstration the planar MX_3 molecule shown in Figure 4.5 and which belongs to point group D_{3h} will be chosen as the example.

A transformation table which details the fate of each hybrid orbital under the influence of each symmetry operation must be constructed. The

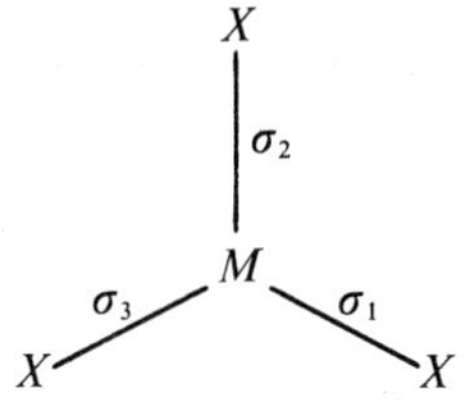

Figure 4.5
Planar MX_3

D_{3h}	E	C_3^1	C_3^2	C_2^1	C_2^2	C_2^3	σ_h	S_3^1	S_3^2	σ_v^1	σ_v^2	σ_v^3
σ_1	σ_1	σ_3	σ_2	σ_1	σ_3	σ_2	σ_1	σ_3	σ_2	σ_1	σ_3	σ_2
σ_2	σ_2	σ_1	σ_3	σ_3	σ_2	σ_1	σ_2	σ_1	σ_3	σ_3	σ_2	σ_1
σ_3	σ_3	σ_2	σ_1	σ_2	σ_1	σ_3	σ_3	σ_2	σ_1	σ_2	σ_1	σ_3

Figure 4.6
Transformation Table For MX_3

transformation table is shown in Figure 4.6. From the considerations outlined in Section 4-3 we see that $\mathbf{\Gamma}_\sigma = 3\ 0\ 1\ 3\ 0\ 1$ and $\mathbf{\Gamma}_\sigma = A_1' + E'$. Symmetry orbitals result from the multiplication of each row in the transformation table by the two irreducible representations. Three symmetry orbitals should result; one A_1' and two E'.

By multiplication of the first row by A_1' there results

$4(\sigma_1 + \sigma_2 + \sigma_3)$ from the first line
$4(\sigma_1 + \sigma_2 + \sigma_3)$ from the second line
$4(\sigma_1 + \sigma_2 + \sigma_3)$ from the third line

The A_1' normalized symmetry orbital is $(\sigma_1 + \sigma_2 + \sigma_3)/\sqrt{3}$.

Multiplication of each line in the transformation table by the E' irreducible representation yields:

$2(2\sigma_1 - \sigma_2 - \sigma_3)$ from the first line
$2(2\sigma_2 - \sigma_3 - \sigma_1)$ from the second line
$2(2\sigma_3 - \sigma_2 - \sigma_1)$ from the third line

These three expressions are not mutually orthogonal. In order to make them orthogonal we will choose one, say $2\sigma_1 - \sigma_2 - \sigma_3$, and take a combination of the other two. Here we subtract the third combination from the second function to eliminate σ_1. The normalized symmetry orbitals are $(1/\sqrt{6})(2\sigma_1 - \sigma_3 - \sigma_2)$ and $(1/\sqrt{2})(\sigma_2 - \sigma_3)$.

In matrix notation the results may be summarized as

$$\begin{pmatrix}\sigma_1'\\ \sigma_2'\\ \sigma_3'\end{pmatrix} = \begin{pmatrix}1/\sqrt{3} & 1/\sqrt{3} & 1/\sqrt{3}\\ 2/\sqrt{6} & -1/\sqrt{6} & -1/\sqrt{6}\\ 0 & 1/\sqrt{2} & -1/\sqrt{2}\end{pmatrix}\begin{pmatrix}\sigma_1\\ \sigma_2\\ \sigma_3\end{pmatrix} \tag{4-6}$$

where $\sigma_1', \sigma_2', \sigma_3'$ are a set of functions that have the symmetry of the group and $\sigma_1, \sigma_2, \sigma_3$ are the hybrid orbitals. A set of functions $\sigma_1', \sigma_2', \sigma_3'$ which have the required symmetry are the atomic orbitals which transform as

$A'_1 + E'$ or the s and (p_x, p_y) orbitals. By pre-multiplication (that is multiplication on the left) of each side of Equation (4-6) by the inverse of the 3×3 matrix, the expressions of the hybrid orbitals $\sigma_1, \sigma_2, \sigma_3$ in terms of the symmetry orbitals of the group $\sigma'_1, \sigma'_2, \sigma'_3$ result. Since we are dealing with orthonormal functions, the transpose of the matrix is equal to its inverse. Thus

$$\begin{pmatrix} 1/\sqrt{3} & 2/\sqrt{6} & 0 \\ 1/\sqrt{3} & -1/\sqrt{6} & 1/\sqrt{2} \\ 1/\sqrt{3} & -1/\sqrt{6} & -1/\sqrt{2} \end{pmatrix} \begin{pmatrix} 1/\sqrt{3} & 1/\sqrt{3} & 1/\sqrt{3} \\ 2/\sqrt{6} & -1/\sqrt{6} & -1/\sqrt{6} \\ 0 & 1/\sqrt{2} & -1/\sqrt{2} \end{pmatrix} \begin{pmatrix} \sigma_1 \\ \sigma_2 \\ \sigma_3 \end{pmatrix}$$
$$= \begin{pmatrix} 1/\sqrt{3} & 2/\sqrt{6} & 0 \\ 1/\sqrt{3} & -1/\sqrt{6} & 1/\sqrt{2} \\ 1/\sqrt{3} & -1/\sqrt{6} & -1/\sqrt{2} \end{pmatrix} \begin{pmatrix} \sigma'_1 \\ \sigma'_2 \\ \sigma'_3 \end{pmatrix}$$

and

$$\begin{pmatrix} \sigma_1 \\ \sigma_2 \\ \sigma_3 \end{pmatrix} = \begin{pmatrix} 1/\sqrt{3} & 2/\sqrt{6} & 0 \\ 1/\sqrt{3} & -1/\sqrt{6} & 1/\sqrt{2} \\ 1/\sqrt{3} & -1/\sqrt{6} & -1/\sqrt{2} \end{pmatrix} \begin{pmatrix} s \\ p_x \\ p_y \end{pmatrix}$$

Therefore, the three analytical functions are

$$\sigma_1 = (1/\sqrt{3})s + (2/\sqrt{6})p_x$$
$$\sigma_2 = (1/\sqrt{3})s - (1/\sqrt{6})p_x + (1/\sqrt{2})p_y$$
$$\sigma_3 = (1/\sqrt{3})s - (1/\sqrt{6})p_x - (1/\sqrt{2})p_y$$

The same procedure may be used for any other geometry although the algebra increases considerably as the symmetry becomes higher and the number of operations increases.

Exercise

4-4: Generate the appropriate linear combination of atomic orbitals which form σ-hybrid orbitals for central atoms in molecules with the following geometries: a) square planar, b) tetrahedral, c) octahedral.

4-5 PI BONDING IN A TETRAHEDRAL MOLECULE

To illustrate the technique used to determine the π-orbitals on the ligand atoms, which may combine with the remaining orbitals on the central atom, we will use the coordinate system shown in Figure 4.7. In Figure

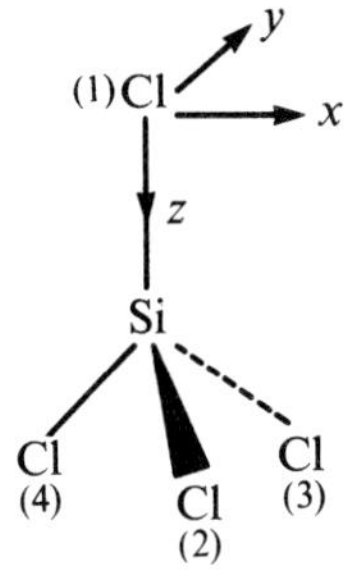

Note: $SiCl_4$ is tetrahedral. The x-axis on Cl(1) is in the plane of the paper and the y-axis points into the page.

Figure 4.7
Ligand Atom Pi Orbitals

4.7 the chlorine p_z orbital is pointed toward the silicon atom and is used in sigma bonding. The remaining p orbitals are designated by x and y. Here it is assumed that the chlorine 3s orbital is not used in bonding since it is energetically much more stable than the p orbitals.

To construct the reducible representation spanned by these 8 chlorine p orbitals, we will use the technique of counting those bonds not shifted by the operation. Thus for E the character is 8, for C_2 and S_4 $\chi = 0$ since all chlorines are shifted by these operations and for σ_d $\chi = 0$ because for each of the two chlorine atoms in the σ_d plane one orbital is unaffected and contributes $+1$ to the character while the other orbital is converted to the negative of itself and contributes -1 to the character; the sum is therefore zero. This point is illustrated by the diagrams in Figure 4.8. Here we see that the σ_d operation interchanges chlorine atoms 3 and 2, and

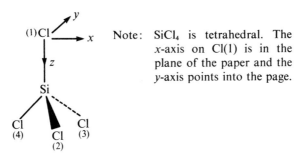

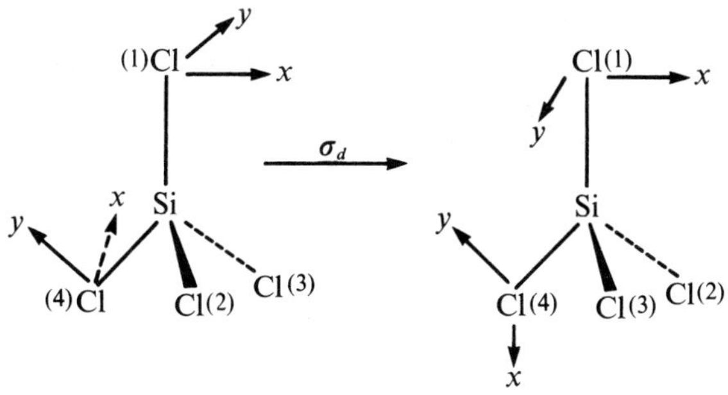

Figure 4.8
σ_d Operation on Pi Orbitals

they contribute nothing to the character. Also, we see that

$$x_1 \longrightarrow x_1 \quad \text{contributes } +1 \text{ to the character}$$
$$y_1 \longrightarrow -y_1 \quad \text{contributes } -1 \text{ to the character}$$
$$x_4 \longrightarrow -x_4 \quad \text{contributes } -1 \text{ to the character}$$
$$y_4 \longrightarrow y_4 \quad \text{contributes } +1 \text{ to the character}$$

therefore $\chi = 0$.

A different problem arises for the C_3 operation. After the C_3 operation, say along the Si-Cl_1 bond, chlorine atoms 2, 3, and 4 are interchanged and need not be considered further. However, the new orbitals x' and y' which result from the C_3 operation on x_1 and y_1 must be described in terms of the old set x_1 and y_1. We use the transformation matrix

$$\begin{pmatrix} \cos\theta & \sin\theta \\ \sin\theta & \cos\theta \end{pmatrix}$$

for $\theta = 120°$, and find the character for C_3 to be -1.

Thus the reducible representation and its components are

T_d	E	$8C_3$	$3C_2$	$6S_4$	$6\sigma_d$	
Γ_π	8	-1	0	0	0	$= E + T_1 + T_2$

By consulting the character table we find that the set $(d_{x^2-y^2}, d_{z^2})$ transforms as E, the sets (p_x, p_y, p_z) and (d_{xz}, d_{yz}, d_{xy}) transform as T_2, and that there are no central atom orbitals which transform as T_1. This means that of the eight possible π-bonds (as determined by the number of ligand π-orbitals) only five can form because suitable orbitals are available for only five. It was noted above that the atomic orbitals used in the formation of sigma bonds would include a combination of the two sets which transform as T_2. The remaining set of three orbitals may be used to form the π-bonds. The T_1 combinations will be nonbonding.

Exercises

4–5: Describe the symmetry aspects of π-bonding in the molecule AB_2C_4 which belongs to the point group D_{4h}. Assume A is a transition metal with $Z < 30$ and that B and C are donor atoms with $Z < 10$.

4–6: Compare π-bonding in BF_3 and NO_3^-.

4–7: What do the following figures actually illustrate?

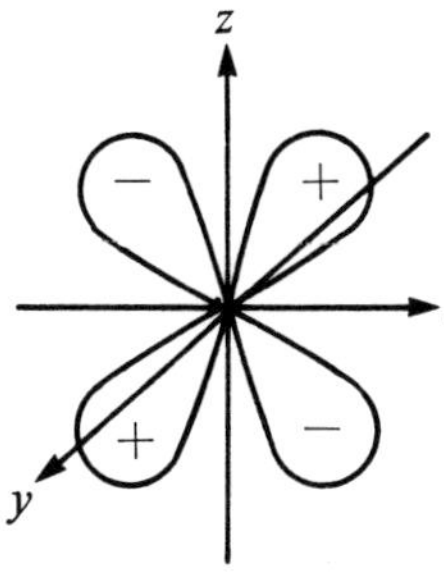

d^2_{xz}

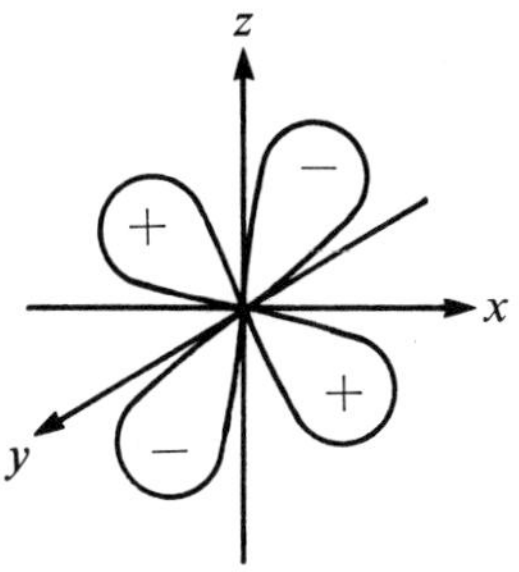

d^2_{yz}

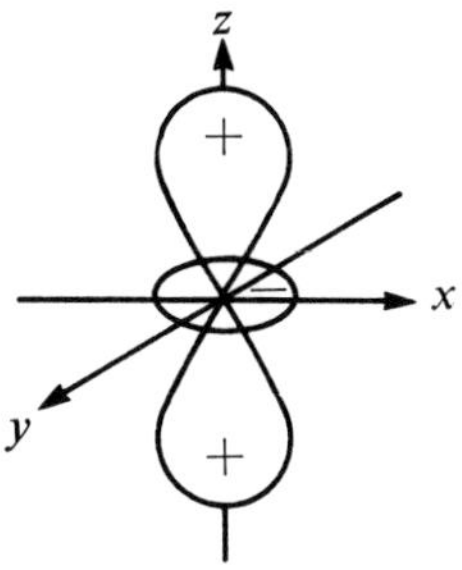

$d^2_{z^2}$

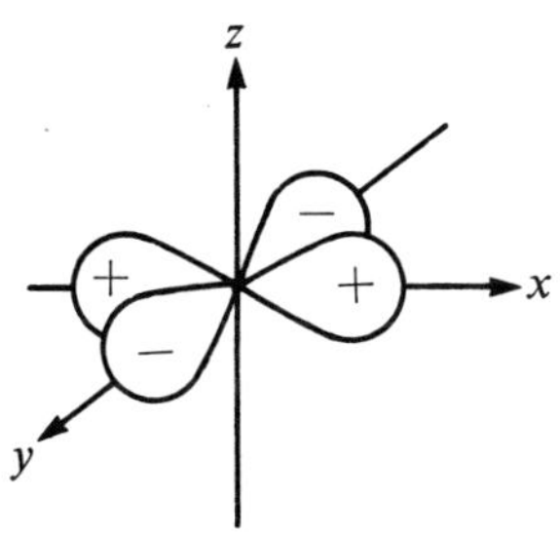

$d^2_{x^2-y^2}$

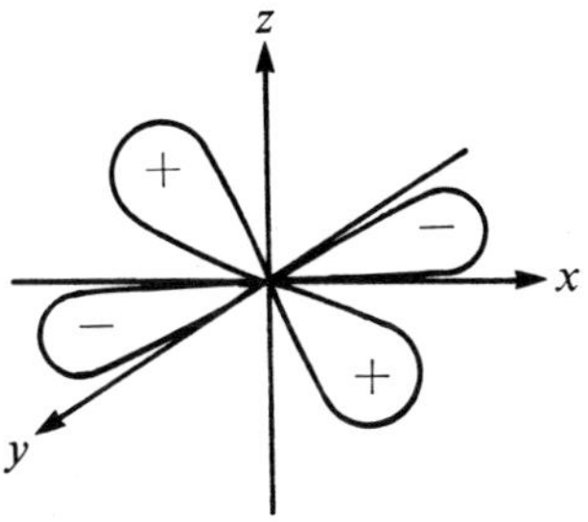

d^2_{xy}

Bibliography

D. S. Urch, *Orbitals and Symmetry.* Harmondsworth, Middlesex, England: Penguin Books, 1970.

G. Herzberg, *Atomic Spectra and Atomic Structure.* New York: Dover Publications, 1944.

5

VIBRATIONAL SPECTROSCOPY

Vibrational motions of molecules may be excited by the absorption of energy, largely in the range 50 to 4,000 cm^{-1}. While detailed studies of the vibrational spectra of simple molecules can yield much information about chemical bonds and molecular structure, simple comparisons of spectral features for series of complicated molecules are useful for characterization purposes. Although the theoretical basis for the interpretation of vibrational spectra has been known for quite some time, there is currently much interest and research in vibrational spectroscopy, mainly due to the recent availability of Raman spectrometers using laser sources and to the complementary advances in infrared instrumentation.

Concepts and techniques of group theory are very useful for the understanding and systematization of vibrational spectra. In this chapter we will present some of the elementary theoretical considerations necessary for the interpretation and use of vibrational spectra. These techniques will be illustrated by application to some examples. The material in Sections 5-10 through 5-13 may be skipped by those who are primarily interested in application of symmetry concepts for structural assignments from selection rules.

5-1 HOOKE'S LAW

Consider the simple ball and spring system in Figure 5.1. According to Hooke's law the force on the ball tending to restore it to the equilibrium position is proportional to the displacement, x, of the ball and is given by

$$f = -kx \tag{5-1}$$

where k is the spring stiffness or force constant. The differential change in potential energy of the ball is given by the force that the spring exerts on the ball times the differential change in distance, or

$$dV = -f\,dx \tag{5-2}$$

Thus,

$$dV = kx\,dx$$

which can be integrated to give

$$V = \tfrac{1}{2}kx^2 \tag{5-3}$$

where the equilibrium position is taken to have zero potential energy.

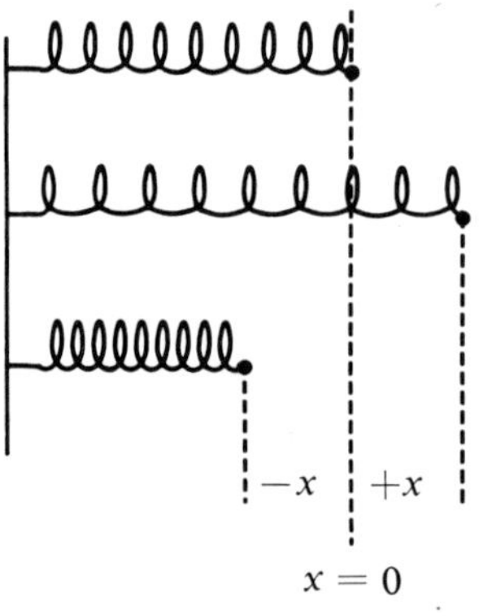

Figure 5.1
(a) The Spring at Rest with the Ball at the Equilibrium Position. (b) The Spring Extended with the Ball Displaced by the Distance x or (c) −x from the Equilibrium Position

Equation (5-3) says that the potential energy increases parabolically as the ball moves in either direction from the equilibrium position, as is shown in Figure 5.2. This system corresponds to a harmonic oscillator.

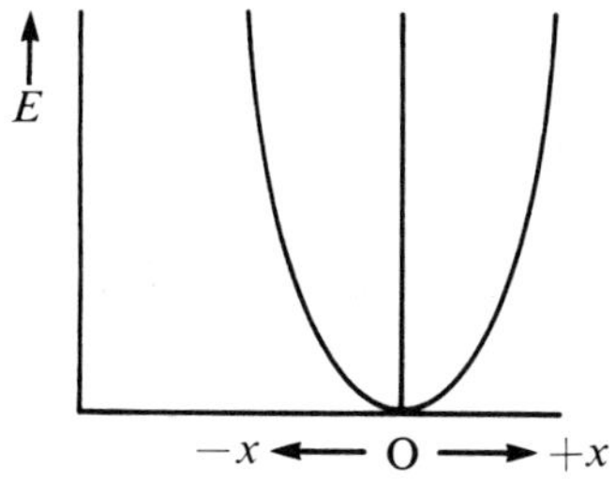

Figure 5.2
The Potential Energy of the Ball and Spring System

The ball and spring model, which allows a continuum of energy states for the vibrating system, is not applicable to molecular vibrations. From quantum mechanics only certain energy states are possible for molecular vibrations, as indicated in Figure 5.3. The vibrational energy states are quantized and their relative energies are given by

$$E_v = h\nu(v + \tfrac{1}{2}) \tag{5-4}$$

where $v = 0, 1, 2, 3, \ldots$.

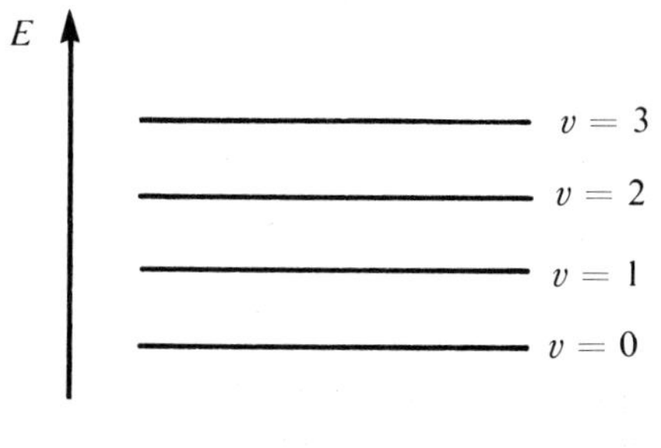

Figure 5.3
Vibrational Energy States for a Diatomic Molecule

5-2 MODEL FOR DIATOMIC MOLECULES

A mechanical model for a diatomic molecule is provided in Figure 5.4 by two particles of mass m_1 and m_2 connected by a spring. If the two particles are allowed to move along the interparticle line only, then from Hooke's law

$$V = \tfrac{1}{2}k(x_2 - x_1)^2$$

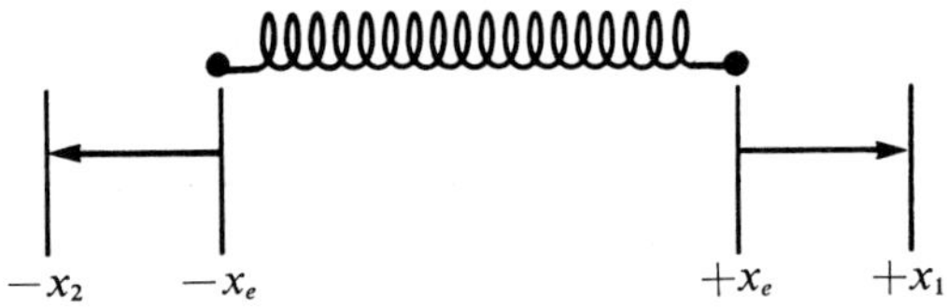

Figure 5.4
Displacement of Two Particles Connected by a Spring

The kinetic energy associated with this motion is

$$T = \tfrac{1}{2}m_1\dot{x}_1^2 + \tfrac{1}{2}m_2\dot{x}_2^2$$

Applying Lagrange's equation, which is given in Equation (5-5),

$$\frac{d}{dt}\left(\frac{\partial T}{\partial \dot{x}_i}\right) + \frac{\partial V}{\partial x_i} = 0 \tag{5-5}$$

we can write the differential equations

$$\begin{aligned} m_1\ddot{x}_1 - k(x_2 - x_1) &= 0 \\ m_2\ddot{x}_2 + k(x_2 - x_1) &= 0 \end{aligned} \tag{5-6}$$

To solve these equations we try the solutions

$$\begin{aligned} x_1 &= A_1 \cos(2\pi\nu t + \phi) \\ x_2 &= A_2 \cos(2\pi\nu t + \phi) \end{aligned} \tag{5-7}$$

where ν is the frequency of the motion. Taking the second derivatives of x_1 and x_2 (Equation 5-7) and substituting into Equation (5-6) yields

$$\begin{aligned} (-4\pi^2\nu^2 m_1 + k)A_1 - kA_2 &= 0 \\ -kA_1 + (-4\pi^2\nu^2 m_2 + k)A_2 &= 0 \end{aligned} \tag{5-8}$$

In the usual manner, if the determinant of the coefficients of A_1 and A_2 is set equal to zero, the result is

$$\nu^2 = \frac{k(m_1 + m_2)}{4\pi^2 m_1 m_2}$$

The factor $m_1m_2/(m_1 + m_2)$ is the reduced mass, which is designated by μ, so we may write

$$\nu = \frac{1}{2\pi}\sqrt{\frac{k}{\mu}} \qquad \textbf{(5-9)}$$

With this fundamental equation one can calculate the frequency of vibration if the reduced mass and force constant are known. Usually, however, the equation is used to calculate force constants from experimentally determined frequencies.

It is appropriate to mention units at this point. Most frequently, vibrational bands are expressed in terms of wave numbers, not frequencies, although it is not uncommon to see the statement: "the frequency of the vibration is ____ cm^{-1}." Of course, frequency with units of sec^{-1} is converted to ν (wave numbers) by dividing by the speed of light. Force constants are usually expressed as millidynes/A, or 10^5 dynes/cm.

5-3 ANHARMONICITY OF MOLECULAR VIBRATIONS

Equation (5-4), which is based on the harmonic oscillator model, requires that the vibrational energy levels be equally spaced. However, the harmonic oscillator model is only an approximation, and the energy levels in real molecules usually become closer together as the vibrational quantum number v increases. Consequently, the energy of the transition from the energy state with $v = 0$ to that with $v = 2$ may be slightly less than twice the energy of the transition from the state with $v = 0$ to that with $v = 1$. The transition from $v = 0$ to $v = 1$ is called the fundamental and that from $v = 0$ to $v = 2$ the first overtone. For HCl the fundamental occurs at 2,885.9 cm^{-1}, and the first overtone at 5,668.0, the second overtone at 8,347.0 cm^{-1} and the third at 10,923.1. Were HCl an harmonic oscillator the third overtone would have occurred at 11,543.6 cm^{-1}. There is no reason to require the anharmonicity coefficient to be negative. Data for the azide ion have been interpreted in terms of a positive anharmonicity coefficient. In other words, the first overtone occurs at an energy more than twice that of the fundamental.

5-4 NORMAL COORDINATES

The vibrational motions of molecules are usually described in terms of a normal coordinate analysis. The nature of normal coordinates is best described by an illustration. Consider a body of mass m held by springs as shown in Figure 5.5. For the discussion assume that Hooke's law holds and that the spring force constants are k_x and k_y. Since the component of the force in a given direction is given by the magnitude of the force times the cosine of the angle between the direction of the force and the direction being considered, then for small displacements in the x-direction the restoring force in the x-direction arises from a displacement in the x-direction. A similar situation holds for the y-direction.

Thus

$$f_x = -\left(\frac{\partial V}{\partial x}\right)_y = -k_x x$$

and

$$f_y = -\left(\frac{\partial V}{\partial y}\right)_x = -k_y y$$

permitting us to write

$$V = \tfrac{1}{2}k_x x^2 + \tfrac{1}{2}k_y y^2$$

The kinetic energy may be written as

$$T = \tfrac{1}{2}m\dot{x}^2 + \tfrac{1}{2}m\dot{y}^2$$

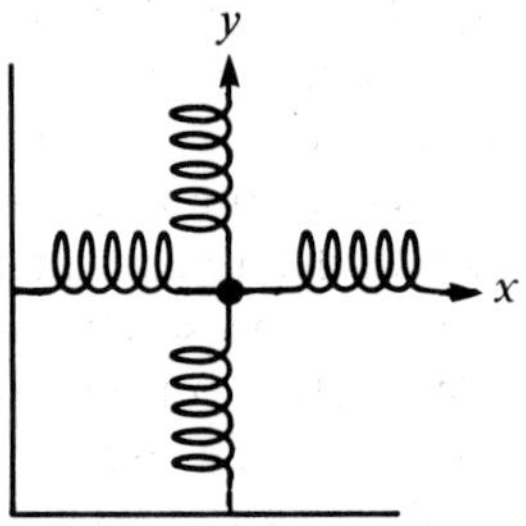

Figure 5.5
Spring System to Illustrate Normal Coordinates

Now apply Lagrange's equations; there results

$$m\ddot{x} + k_x x = 0$$

and

$$m\ddot{y} + k_y y = 0$$

which have the solutions

$$x = A_x \sin(2\pi\nu t + \phi) \qquad \textbf{(5-11a)}$$

$$y = A_y \sin(2\pi\nu t + \phi) \qquad \textbf{(5-11b)}$$

Thus, initial displacement along the x or y direction will lead to simple sinusoidal vibration along that axis, but displacement along any other di- ection will lead to a motion which must be described in terms of components along x and y. Since the x and y coordinates allow the motion to be described most simply, these coordinates are called normal coordinates, and the motion along these coordinates is called a normal mode.

5-5 GROUP THEORY CONSIDERATIONS

For a linear molecule containing N atoms, there are $3N$-5 fundamental molecular vibrations and for a nonlinear molecule, $3N$-6. Each of the normal modes of vibration forms the basis for an irreducible representation of the point group to which the molecule belongs. All the atomic motions in a molecule can be described in terms of sets of basis vectors situated on the individual atoms. Construction of the reducible representation, $\Gamma_{\text{b.v.}}$, spanned by basis vectors gives the symmetry properties of all the molecular motions including rotations and translations. In each point group rotations transform as R_x, R_y, and R_z and translations as x, y, and z. Subtracting them from the representation $\Gamma_{\text{b.v.}}$ leaves the symmetry species to which the fundamental modes of vibration belong.

The contributions of the various atomic motions to the fundamental modes may be determined by using certain features of molecules as basis sets for the representations. For example, symmetries of the stretching vibrations may be determined by using the bonds themselves as a representation, and bending vibrations may be represented by the angles between the chemical bonds which form the angles. These features are called internal coordinates. A stretching motion may couple with bending motions if the motions transform as the same irreducible representation of the point group.

The procedure can be illustrated by analyzing the motions of the H_2O molecule, which is a non-linear three-atom molecule with three fundamental modes. The molecule belongs to the point group C_{2v}.

Step 1.) Draw a model with a set of basis vectors on each atom. Note that in Figure 5.6 all similarly labeled vectors are drawn parallel to each other.

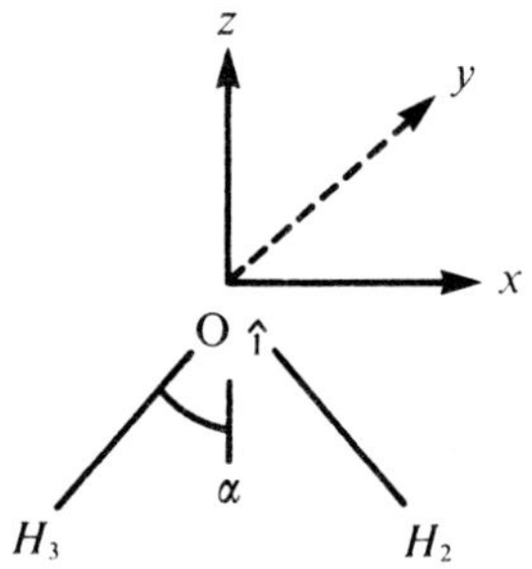

Figure 5.6
Basis Vectors for Water Molecule

Step 2.) Construct the reducible representation spanned by the nine basis vectors. The point group C_{2v} includes the symmetry operations E, C_2, $\sigma_v(XZ)$, and $\sigma_v(YZ)$ for which the representation matrices may be written in the manner described in Chapter 2.

The characters of the representation matrices form the reducible representation, giving,

C_{2v}	E	C_2	$\sigma(xz)$	$\sigma(xy)$
$\Gamma_{\text{b.v.}}$	9	-1	3	1

Step 3.) Decompose the reducible representation to yield $\Gamma_{\text{b.v.}} = 3A_1 + A_2 + 3B_1 + 2B_2$.

Step 4.) Subtract rotations and translations. In C_{2v} the transformation properties are:

R_z transforms as A_2
R_y transforms as B_1
R_x transforms as B_2
T_z transforms as A_1
T_y transforms as B_2
T_x transforms as B_1

Subtraction of $\Gamma_{\text{translations}}$ and $\Gamma_{\text{rotations}}$ leaves $\Gamma_{\text{vib}} = 2A_1 + B_1$.

Step 5.) Determine the contributions of the bond stretching vibrations and angle deformations by examining the symmetry properties of the internal coordinates. The reducible representation spanned by the stretching motions of a set of bonds may be constructed by counting the number of bonds not shifted during the operation. These manipulations allow the following reducible representations to be written:

C_{2v}	E	C_2	$\sigma_v(xz)$	$\sigma_v(yz)$	
$\Gamma_{\text{O-H bonds}}$	2	0	2	0	$= A_1 + B_1$
$\Gamma_{\text{H-O-H angle}}$	1	1	1	1	$= A_1$

The two A_1 modes include symmetric bond stretching and angle deformation, and the B_1 mode includes only antisymmetric bond stretching.

For additional practice we will work out the contributions of the atomic motions to the fundamental motions of the trigonal planar molecule BF_3. The basis vectors are shown in Figure 5.7. The 12 basis vectors span the representation:

D_{3h}	E	$2C_3$	$3C_2$	σ_h	$2S_3$	$3\sigma_v$
$\Gamma_{\text{b.v.}}$	12	0	-2	4	-2	2

which may be decomposed to yield

$$\Gamma_{\text{b.v.}} = A_1' + A_2' + 3E' + 2A_2'' + E''$$

From the D_{3h} character table we see that

$$\Gamma_{\text{trans}} = A_2'' + E'$$

and

$$\Gamma_{\text{rot}} = A_2' + E''$$

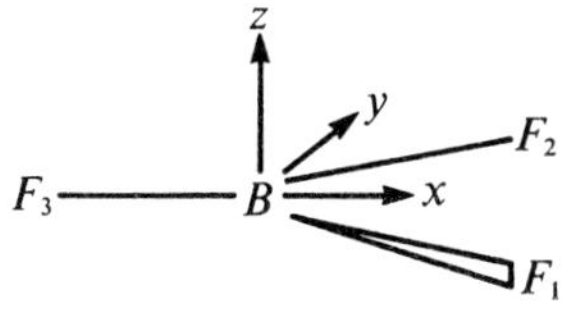

Figure 5.7
Coordinate System for BF_3.

Thus,

$$\Gamma_{\text{vib}} = A_1' + 2E' + A_2''$$

In other words, of the six fundamental motions, one forms a basis for an A_1' irreducible representation, another for an A_2'' irreducible representation, while two sets of two each form bases for E' irreducible representations. The A_1' and A_2'' fundamental modes are singly degenerate while the E' stands for doubly degenerate fundamental modes.

Now we will analyze the contributions of the internal coordinates to the fundamental motions. First choose the three bond stretching coordinates, for which we can construct the representation

D_{3h}	E	$2C_3$	$3C_2$	σ_h	$2S_3$	$3\sigma_v$	
$\Gamma_{3\text{ bonds}}$	3	0	1	3	0	1	$= A_1' + E'$

Thus, the symmetric stretching motion of all three *B-F* bonds makes up the A_1' fundamental mode, while a set of two antisymmetric motions makes up the E'.

Next choose the set of three *F-B-F* angles (in the plane) as a basis. Here we obtain

$$\Gamma_{3(F\text{-}B\text{-}F)\text{ angles}} = A_1' + E'$$

A little thought reveals that all three of the *F-B-F* angles can not increase or decrease simultaneously by the same amount. This situation arises because the three angles are not independent. Therefore, the A_1' coordinate is discarded; it is called the *redundant species.* Redundancies arise especially for planar, tetrahedral, and octahedral molecules. They are usually easily spotted.

The E' vibrations therefore are antisymmetric angle deformations. It is important to notice that both bond stretching motions and angle deformations form a basis for the E' irreducible representation. The physical consequence of this is important. Since the motions have the same symmetry, they can mix, and it is likely that the observed fundamental motions of E' symmetry will include contributions from angle deformations and bond stretching vibrations.

At this point we have accounted for five of the normal modes. The sixth normal mode, of A_2'' symmetry, is the out-of-plane deformation. It is easy to show that the out-of-plane deformation forms a basis for A_2''.

There is a bit of a problem in the determination of the irreducible representations to which the internal coordinates of linear molecules belong. Since linear molecules belong to the infinite point groups $D_{\infty h}$ and $C_{\infty v}$, the technique used in Chapter 2 fails. However, it is possible to determine the irreducible representations by a descent in symmetry method*. This method is based on the fact that the basis of a representation of a group G is also the basis of a representation of another group G_s, which is a subgroup of the group G. For the method to work the z axis must be taken colinear with the C_∞ axis of the parent group.

Consider the linear molecule $O = C = S$ which belongs to the point group $C_{\infty v}$. Now choose the subgroup C_{2v} and construct the representation spanned by the set of basis vectors. The representation is

	E	C_2	σ_v^{xz}	σ_v^{yz}
$\Gamma_{\text{b.v.}}$	9	-3	3	3

which may be decomposed to yield $3A_1 + 3B_1 + 3B_2$. Subtracting $\Gamma_{\text{trans}} + \Gamma_{\text{rot}} = A_1 + 2B_1 + 2B_2$, we get $\Gamma_{\text{vib}} = 2A_1 + B_1 + B_2$. Note that rotation about the z axis is not a valid rotational degree of freedom for linear molecules. Now compare the bases of the irreducible representations of $C_{\infty v}$ and C_{2v}:

C_{2v}	basis	$C_{\infty v}$
A_1	z	Σ^+
B_1	x	Π
B_2	y	

It is clear that the two A_1 irreducible representations correlate with Σ^+, and that the two representations B_1 and B_2 correlate with Π. Thus, two of the fundamental modes of vibration of the $O = C = S$ molecule, or any linear triatomic molecule belonging to the group $C_{\infty v}$, have symmetry Σ^+, while a set of the other two fundamental modes transform as Π and are doubly degenerate.

We shall do one more example, that being $O = C = O$ which transforms as $D_{\infty h}$. A convenient subgroup is D_{2h}, and the set of basis vectors forms the reducible representation

D_{2h}	E	$C_2(z)$	$C_2(y)$	$C_2(x)$	i	$\sigma_v(xy)$	$\sigma_v(xz)$	$\sigma_v(yz)$
$\Gamma_{\text{b.v.}}$	9	-3	-1	-1	-3	1	3	3

which may be decomposed to give

$$\Gamma_{\text{b.v.}} = A_g + B_{2g} + B_{3g} + 2B_{1u} + 2B_{2u} + 2B_{3u}$$

*D.P. Strommen and E.R. Lippincott, *J. Chem. Ed.*, *49*, 341 (1972).

Subtracting

$$\Gamma_{\text{rot}} + \Gamma_{\text{trans}} = B_{2g} + B_{3g} + B_{1u} + B_{2u} + B_{3u}$$

we get

$$\Gamma_{\text{vib}} = A_g + B_{1u} + B_{2u} + B_{3u}$$

Now compare the bases of the irreducible representations in $D_{\infty h}$ and D_{2h}:

D_{2h}	basis	$D_{\infty h}$
A_g	z^2	Σ_g^+
B_{1u}	z	Σ_u^+
B_{2u}	x	Π_u
B_{3u}	y	

Two of the four fundamental modes of vibration form a basis for Π_u, while a third transforms as Σ_g^+, and a fourth as Σ_u^+.

Exercise 5-1

For the following molecules determine a) which irreducible representations are spanned by the normal modes of vibration, and b) which internal coordinates contribute to the various fundamental modes:

1. NF_3
2. CH_4
3. $[PtCl_4]^{2-}$ (square planar)
4. $[CuCl_5]^{3-}$ (trigonal bipyramidal)
5. SF_6
6. C_2H_2

5-6 CONSTRUCTION OF SYMMETRY COORDINATES

There are four types of internal coordinates. These are bond stretching, Δr; angle deformation, $\Delta\alpha$; torsional motion, $\Delta\tau$; and out-of-plane bending. The symmetry coordinates, which are linear combinations of internal coordinates, are constructed in the following way:

1. A transformation table for the internal coordinates is constructed.
2. Each row in the transformation table is multiplied by the irreducible representations of the normal modes.
3. The symmetry coordinates thus constructed are normalized.

As an example consider the pyramidal molecule AB_3 which belongs to point group C_{3v}. From the relationship $3N - 6$, we know that there are 6 vibrational degrees of freedom, and hence six internal coordinates are needed. These are Δr_a, Δr_b, Δr_c, $\Delta\alpha_{ab}$, $\Delta\alpha_{ac}$, and $\Delta\alpha_{bc}$, which transform in the following manner:

C_{3v}	E	$2C_3$	$3\sigma_v$	
$\Gamma_{\Delta r_i}$	3	0	1	$= A_1 + E$
$\Gamma_{\Delta\alpha_{ij}}$	3	0	1	$= A_1 + E$

We must now construct two symmetry coordinates, each of which transforms as A_1, and two sets of two symmetry coordinates with each set forming the basis for an E representation.

The transformation table is given in Table 5.1. Multiplying each row in the transformation table by the A_1 irreducible representation yields for the A_1 symmetry coordinates the linear combinations

$$2(\Delta r_a + \Delta r_b + \Delta r_c)$$

$$2(\Delta\alpha_{ab} + \Delta\alpha_{ac} + \Delta\alpha_{bc})$$

which are normalized to give

$$\left(\frac{1}{\sqrt{3}}\right)(\Delta r_a + \Delta r_b + \Delta r_c)$$

$$\left(\frac{1}{\sqrt{3}}\right)(\Delta\alpha_{ab} + \Delta\alpha_{ac} + \Delta\alpha_{bc})$$

In the same manner multiplication by the characters of the E irreducible representation yields

$$2\Delta r_a - \Delta r_b - \Delta r_c$$

$$2\Delta r_b - \Delta r_c - \Delta r_a$$

$$2\Delta r_c - \Delta r_a - \Delta r_b$$

Table 5.1
Transformation Table

C_{3v}	E	C_3^1	C_3^2	σ_v^a	σ_v^b	σ_v^c
Δr_a	Δr_a	Δr_c	Δr_b	Δr_a	Δr_c	Δr_b
Δr_b	Δr_b	Δr_a	Δr_c	Δr_c	Δr_b	Δr_a
Δr_c	Δr_c	Δr_b	Δr_a	Δr_b	Δr_a	Δr_c
$\Delta\alpha_{ab}$	$\Delta\alpha_{ab}$	$\Delta\alpha_{ac}$	$\Delta\alpha_{bc}$	$\Delta\alpha_{ac}$	$\Delta\alpha_{bc}$	$\Delta\alpha_{ab}$
$\Delta\alpha_{ac}$	$\Delta\alpha_{ac}$	$\Delta\alpha_{bc}$	$\Delta\alpha_{ab}$	$\Delta\alpha_{ab}$	$\Delta\alpha_{ac}$	$\Delta\alpha_{bc}$
$\Delta\alpha_{bc}$	$\Delta\alpha_{bc}$	$\Delta\alpha_{ab}$	$\Delta\alpha_{ac}$	$\Delta\alpha_{bc}$	$\Delta\alpha_{ab}$	$\Delta\alpha_{ac}$

and

$$2\Delta\alpha_{ab} - \Delta\alpha_{bc} - \Delta\alpha_{ac}$$
$$2\Delta\alpha_{ac} - \Delta\alpha_{ab} - \Delta\alpha_{bc}$$
$$2\Delta\alpha_{bc} - \Delta\alpha_{ac} - \Delta\alpha_{ab}$$

Neither of these sets is linearly independent. This is readily seen since the determinant of the coefficients vanishes. Thus, for bond stretching we will choose as one symmetry coordinate

$$2\Delta r_a - \Delta r_b - \Delta r_c$$

and a combination of the remaining two chosen in such a way as to eliminate Δr_a. The normalized symmetry coordinates which result are shown in Equation (5-12).

$$\left(\frac{1}{\sqrt{6}}\right)(2\Delta r_a - \Delta r_b - \Delta r_c) \quad \textbf{(5-12a)}$$

$$\left(\frac{1}{\sqrt{2}}\right)(\Delta r_b - \Delta r_c) \quad \textbf{(5-12b)}$$

An analogous set of symmetry coordinates can be written for the angle deformations. They are

$$\left(\frac{1}{\sqrt{6}}\right)(2\Delta\alpha_{ab} - \Delta\alpha_{ac} - \Delta\alpha_{bc})$$

$$\left(\frac{1}{\sqrt{2}}\right)(\Delta\alpha_{ac} - \Delta\alpha_{bc})$$

The symmetry coordinates are useful in symmetry factoring the $|G$ and $|F$ matrices, which are used in the normal coordinate analysis of the vibrational problem. We define the $|U$ matrix as the transformation matrix which relates the symmetry coordinates to the internal coordinates. Thus, for the AB_3 example, the $|U$ matrix may be seen to be

$$\begin{pmatrix} s_1 \\ s_2 \\ s_3 \\ s_4 \\ s_5 \\ s_6 \end{pmatrix} = \begin{pmatrix} 1/\sqrt{3} & 1/\sqrt{3} & 1/\sqrt{3} & 0 & 0 & 0 \\ 0 & 0 & 0 & 1/\sqrt{3} & 1/\sqrt{3} & 1/\sqrt{3} \\ 2/\sqrt{6} & -1/\sqrt{6} & -1/\sqrt{6} & 0 & 0 & 0 \\ 0 & 0 & 0 & 2/\sqrt{6} & -1/\sqrt{6} & -1/\sqrt{6} \\ 0 & 1/\sqrt{2} & -1/\sqrt{2} & 0 & 0 & 0 \\ 0 & 0 & 0 & 0 & 1/\sqrt{2} & -1/\sqrt{2} \end{pmatrix} \begin{pmatrix} \Delta r_a \\ \Delta r_b \\ \Delta r_c \\ \Delta\alpha_{ab} \\ \Delta\alpha_{ac} \\ \Delta\alpha_{bc} \end{pmatrix}$$

We will use symmetry coordinates and the $|U$ matrix in the discussion of the vibrations of the water molecule.

Exercise 5-2

Construct the symmetry coordinates for an AX_2 molecule that belongs to point group C_{2v}.

5-7 SELECTION RULES

The intensities (I) of infrared transitions are proportional to integrals of the type

$$\int \Psi_{\text{e.s.}} \mu \, \Psi_{\text{g.s.}} \, d\tau \qquad \textbf{(5-13)}$$

where $\Psi_{\text{e.s.}}$ is the wave function of the excited state, μ is the dipole moment operator, and $\Psi_{\text{g.s.}}$ is the wave function of the ground state. The ground vibrational state is always totally symmetric. If the integral is to be greater than zero, the product of the irreducible representation spanned by the excited state and that of the dipole moment operator must contain the totally symmetric irreducible representation. The dipole moment operator transforms as x, y, and z, so that vibrational modes will be infrared active only if they transform under the same irreducible representation as x, y, or z. As an illustration, the normal modes of water are infrared active since z transforms as A_1 and x as B_1.

For normal modes to be Raman active, transition integrals of the form

$$\int \Psi_{\text{e.s.}} P \, \Psi_{\text{g.s.}} \, d\tau \qquad \textbf{(5-14)}$$

must be nonzero. P, which is the polarizability operator, transforms as either a square or binary product of x, y, or z. By analogy with the infrared selection rules the C_{2v} character table indicates that all three normal modes of water are Raman active.

There are complications which arise, especially for solid samples. Frequently, the symmetry of the site in a crystal which is occupied by a molecule or ion may be lower than the expected symmetry based on chemical grounds. For example, the CO_3^{2-} ion is expected to have symmetry D_{3h}, but in the mineral aragonite the carbonate ion sits on a site in the crystal lattice that has symmetry C_s. As a result the totally symmetric stretching vibration, which is infrared inactive in D_{3h}, becomes active, and a weak band arising from this fundamental mode may be seen in the infrared spectrum. A treatment of the selection rules for solid samples requires a knowledge of the crystal structure of the substance and is conveniently treated by the site or factor group techniques. It is appropriate to treat this problem following the introduction to space group symmetry in Chapter 8.

A second selection rule based on the harmonic oscillator model requires that transitions are allowed only if $\Delta v = \pm 1$. This selection rule allows fundamental transitions but not overtones. Actual molecular vibrations are not harmonic oscillations since the potential energy function is not symmetric around the equilibrium position, and this selection rule is partially broken down. Therefore, overtones are observed. However, the intensities of bands arising from vibrations for which $\Delta v \neq \pm 1$ are usually an order of magnitude lower than those of allowed transitions.

If the overtones arise from singly degenerate fundamental modes, then the irreducible representation to which the excited states belong can be summarized in the following way. All odd overtones belong to the totally symmetric irreducible representation of the point group, and all even overtones have the symmetry of the fundamental mode.

The selection rules for the overtones of degenerate fundamentals are more difficult to work out. Wilson, et al.,* give the following recipes for the determination of the symmetries of the excited states:

doubly degenerate fundamentals

$$\chi_v(R) = \tfrac{1}{2}[\chi(R)\chi_{v-1}(R) + \chi(R^v)] \qquad \textbf{(5-15)}$$

triply degenerate fundamentals

$$\chi_v(R) = \tfrac{1}{3}[2\chi(R)\chi_{v-1}(R) + \tfrac{1}{2}\{\chi(R^2) - [\chi(R)]^2\}\chi_{v-2}(R) + \chi(R^v)] \qquad \textbf{(5-16)}$$

where $\chi_v(R)$ is the character of the operation R for the $(v-1)$th overtone
$\chi(R)$ is the character of the operation R for the fundamental
$\chi(R^v)$ is the character of the R^v operation

Also, $\chi_0(R) = 1$, and $\chi_{-1}(R) = 0$

We illustrate the application of these rules by working out the symmetry species of the second overtone of an E mode for a molecule which belongs to the C_{3v} point group. Here we must determine $\chi(R^3)$ and $\chi_2(R)$. In C_{3v} we have the operations

$$E \qquad 2C_3 \qquad 3\sigma_v$$

Thus, the corresponding R^2 operations are

$$E \qquad C_3^2 \qquad E$$

*E. B. Wilson, J. C. Decius, and P. C. Cross, *Molecular Vibrations* (New York: McGraw-Hill, 1955).

and R^3 operations become

$$E \quad E \quad \sigma_v$$

We can see from the character table that

$$\Gamma(E) = 2 \quad -1 \quad 0$$

and it follows that

$$\Gamma(R^2) \quad 2 \quad -1 \quad 2$$
$$\Gamma(R^3) \quad 2 \quad 2 \quad 0$$

We first calculate $\chi_2(R)$ from the Equation

$$\chi_2(R) = \tfrac{1}{2}[\chi(R)\chi(R) + \chi(R^2)]$$
$$= 3 \quad 0 \quad 1$$

which follows logically from Equation (5-15). Substitution into Equation (5-16) yields

$$\chi_3(R) = 4 \quad 1 \quad 0$$

which may be decomposed to yield the symmetry species $A_1 + A_2 + E$. Therefore, the second overtone of the E fundamental mode gives rise to three excited states which have the symmetries A_1, A_2, and E.

Absorptions which arise from the simultaneous change of two or more vibrational quantum numbers are called combination bands. The representation spanned by the excited state is determined by taking the direct product of the representations of the constituent fundamentals.

Exercise 5-3

For the five molecules in Exercise 5-1, determine a.) which fundamental modes of vibration are infrared active, and b.) which fundamental modes are Raman active.

5-8 THE ν_n SYMBOLISM FOR LABELING VIBRATIONAL MODES

The totally symmetric vibrational mode of highest frequency is labeled ν_1. For linear triatomic molecules the bending mode is labeled ν_2. After all n symmetric modes have been assigned in order of decreasing frequency, the highest frequency antisymmetric vibrational mode is counted next and is labeled ν_{n+1}. The remaining antisymmetric modes are assigned symbols in order of decreasing frequency. For linear triatomic molecules, ν_1 is obviously the symmetrical stretching vibration, and ν_2, the bending mode, is doubly degenerate.

5-9 ANALYSIS OF AN INFRARED SPECTRUM

In the previous sections some basic ideas about molecular vibrations have been presented. It must be emphasized that the analysis of an infrared spectrum can be an extremely complicated operation. For example, an infrared spectrum can frequently contain more bands than there are normal modes of vibration. Consider the spectrum of SO_2 in the gas state. There are seven bands, which occur at 519, 606, 1151, 1361, 1871, 2305, and 2499 cm^{-1}, whereas by the 3N-6 rule there are only three fundamental modes. The bands at 1151, 519 and 1361 cm^{-1} have been assigned to the fundamental modes ν_1, ν_2, and ν_3, respectively.

5-10 FORCE CONSTANTS

In Section 5-2 we introduced force constants in the discussion of the potential energy of the ball and spring system. The analogy between the stiffness of the spring in that mechanical model and some property of the attractive and repulsive forces between two atoms is apparent, and it is of interest to extract force constants from vibrational spectra so that these forces may be compared. The problem reduces to one of writing the appropriate potential and kinetic energy expressions and solving the equation of motion.

Assuming harmonic oscillation the potential energy expression may be written as

$$V = \tfrac{1}{2} \sum_{i,j=1}^{n} k_{ij}(\Delta q_i)(\Delta q_j) \qquad \textbf{(5-17)}$$

where Δq_i represents the displacement coordinates.
Thus, for the water molecule

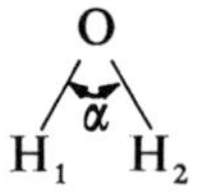

the complete expression is

$$\begin{aligned} V = \tfrac{1}{2}\{&k_{H^1O}(\Delta q_{H^1O})^2 + k_{H^2O}(\Delta q_{H^2O})^2 + k_{H^1H^2}(\Delta q_{H^1H^2})^2 \\ &+ k_{H^1O,H^2O}(\Delta q_{H^1O})(\Delta q_{H^2O}) + k_{H^1O,H^1H^2}(\Delta q_{H^1O})(\Delta q_{H^1H^2}) \\ &+ k_{H^2O,H^1H^2}(\Delta q_{H^2O})(\Delta q_{H^1H^2})\} \end{aligned} \qquad \textbf{(5-18)}$$

Even if we make the logical assumption that $k_{\mathrm{H^1O}} = k_{\mathrm{H^2O}}$, there are four unknown force constants, those being

$$k_{\mathrm{H^1O}} = k_{\mathrm{H^2O}} = k_{\mathrm{HO}}$$

$$k_{\mathrm{H^1H^2}}$$

$$k_{\mathrm{H^1O,H^2O}}$$

$$k_{\mathrm{HO,H^1H^2}}$$

As a general rule antisymmetric stretching vibrations occur at higher energies than symmetric modes, and stretching modes at higher energies than bending modes. The first overtone of ν_1 occurs at 2305 cm^{-1}, and overtones of the other two are not observed. The band at 606 cm^{-1} arises because of a transition from the excited state ν_2 to the excited state ν_1 (Note: $1151 - 519 = 632$ cm^{-1}) which is called a difference band. The remaining two bands result from a simultaneous excitation of two fundamentals; the band at 1871 is assigned to $\nu_2 + \nu_3$ (519 cm^{-1} + 1361 cm^{-1} = 1880 cm^{-1}) and the band at 2499 cm^{-1} to $\nu_1 + \nu_3$ ($1151 + 1361 = 2412$). These latter two are called combination bands.

When the fundamental of an infrared active normal mode occurs at approximately the same energy as an overtone of another normal mode or at the same energy of some combination band, the more intense absorption band will not always merely mask the low intensity band. If the symmetries of the excited states are the same, the fundamental and overtone or combination bands may interact, a phenomenon known as Fermi resonance, and two bands may appear. If there is an exact coincidence of energies, the two bands will be of equal intensity. Neither band can be ascribed to either of the interacting vibrational motions, but to a mixture of the two.

In the analysis of an infrared spectrum Fermi resonance can be detected by isotopic substitution, since vibrational energies are dependent on atomic masses. Isotopic substitution may change the relative energies of the two bands sufficiently so that they no longer occur at the same approximate energy, thus eliminating Fermi resonance. Since there are only three fundamental vibrations, the problem can not be solved uniquely, and it is necessary to express the potential energy in terms of fewer force constants. We could decide to concentrate on changes in internuclear distances—stretchings—and changes in angles between bonds. In this approximation, called the valence bond force field, the potential energy expression becomes

$$V = \tfrac{1}{2}\{k_{\mathrm{HO}}(\Delta_{\mathrm{H^1O}})^2 + k_{\mathrm{HO}}(\Delta_{\mathrm{H^2O}})^2 + k_\alpha(\Delta\alpha)^2\}$$

An additional term

$$k_{\mathrm{HO},\alpha}\{(\Delta_{\mathrm{H^1O}})(\Delta\alpha) + (\Delta_{\mathrm{H^2O}})(\Delta\alpha)\}$$

may be added if data from isotopically substituted species are available. Here the assumption is made that isotopic substitution does not materially affect the force constants. There is much current debate about the best approximate form of the potential energy expression. A consideration of alternate force fields would take us far afield, so we defer the subject by referring the reader to specialized books for a consideration of the Urey-Bradley force field, etc.

Hereafter, force constants will be assigned symbols $f_r, f_{rr}, f_\alpha, f_{\alpha\alpha}, f_{r\alpha}$, etc., where the subscript denotes the type of force constant. The subscripts r and α indicate the force constants for bond stretchings and angle deformations while the double letter subscripts rr, $\alpha\alpha$, and $r\alpha$ indicate interaction force constants. In order for all force constants to have the same units it is necessary to multiply $f_{r\alpha}$ by the bond distance r, which determines Δr, and to multiply f_α and $f_{\alpha\alpha}$ by r^2.

Exercise 5-4

Write out the expression for the potential energy of a molecule AB_3 which belongs to point group C_{3v}. In the generalized valence force field, how many of the force constants must be neglected?

5-11 WILSON F-G MATRIX METHOD

In this method for the solution of the vibrational problem both the kinetic and potential energies are expressed in terms of internal coordinates, which are designated R_k. The description of the method given here follows closely that given by Barrow*. For the potential energy, as we have seen in general, the following expression holds:

$$V = \frac{1}{2} \sum_{j,k} f_{jk} R_j R_k \quad \textbf{(5-19)}$$

It is usually assumed that nonzero terms arise only for those pairs of internal coordinates which have at least one atom in common. In matrix notation the expression for the potential energy becomes

$$2|V = |\boldsymbol{R}'|\boldsymbol{F}|\boldsymbol{R} \quad \textbf{(5-20)}$$

where $|\boldsymbol{R}$ is a column matrix with $3N$-6 rows and $|\boldsymbol{R}'$ is the transpose of $|\boldsymbol{R}$.

---

*G.M. Barrow, *Molecular Spectroscopy* (New York: McGraw-Hill Book Co., 1962), pp. 218–23.

Therefore, $|\boldsymbol{R}'$ is a row matrix with $3N$-6 columns. The $|\boldsymbol{F}$ matrix is the force constant matrix.

We will now construct analogous expressions for the kinetic energy. In terms of Cartesian coordinates the kinetic energy is

$$T = \tfrac{1}{2}\sum_{i=1}^{3n} m_i \dot{x}_i^2$$

If the mass is absorbed into the coordinate by defining $q_i = (m_i)^{1/2} x_i$, then

$$T = \tfrac{1}{2}\sum_{i=1}^{3n} \dot{q}_i^2 \qquad \text{(5-21)}$$

In matrix notation Equation (5-21) becomes

$$2|\boldsymbol{T} = (\dot{\boldsymbol{q}}_i'(\dot{\boldsymbol{q}}_i \qquad \text{(5-22)}$$

We are interested in an expression for the kinetic energy in terms of the internal coordinates R_k. It is clear that the internal coordinates are related to the Cartesian coordinates by geometrical factors, so we can write

$$\dot{R}_k = \sum_{i=1}^{3n} B_{ki}\dot{x}_i$$
$$= \sum_{i=1}^{3n} D_{ki}\dot{q}_i$$

where $D_{ki} = B_{ki}/(m_i)^{1/2}$. In matrix notation the equations may be expressed as

$$|\dot{\boldsymbol{R}} = |\boldsymbol{B}'\dot{\boldsymbol{x}}$$
$$= |\boldsymbol{D}(\dot{\boldsymbol{q}}$$

where $|\dot{\boldsymbol{R}}$ is a column matrix with $3N$-6 rows, $(\dot{\boldsymbol{q}}$ is a column matrix with $3n$ rows, and $|\boldsymbol{B}$ is the transformation matrix containing the geometrical factors. $|\boldsymbol{B}$ has $3n$-6 rows and $3n$ columns.

The $|\boldsymbol{D}$ matrix, which is related to the $|\boldsymbol{B}$ matrix, also has $3n$-6 rows and $3n$ columns. The $|\boldsymbol{D}$ matrix is supplemented with 6 additional rows and made into the square $3n \times 3n$ matrix $|\mathfrak{D}$ by adding the six conditions of zero translational motion and zero angular momentum. This gives

$$|\mathfrak{D} = \begin{pmatrix} |\boldsymbol{D} \\ |\boldsymbol{D}_0 \end{pmatrix}$$

Later on we will need $|\mathfrak{D}^{-1}$. Let it be defined as

$$|\mathfrak{D}^{-1} = [(\boldsymbol{Q}\,(\boldsymbol{Q}_0]$$

where the new matrices $(\boldsymbol{Q}$ and $(\boldsymbol{Q}_0$ will be defined later. However, arbitrarily, $(\boldsymbol{Q}$ is given the dimension $3N \times 3N\text{-}6$ and $(\boldsymbol{Q}_0$ is $3N \times 6$.

We can now write

$$\begin{aligned} \dot{\mathfrak{R}} &= \begin{pmatrix} |\dot{\boldsymbol{R}} \\ |\dot{\boldsymbol{r}} \end{pmatrix} = \begin{pmatrix} |\boldsymbol{D} \\ |\boldsymbol{D}_0 \end{pmatrix} (\dot{\boldsymbol{q}} \\ &= |\boldsymbol{D}\,(\dot{\boldsymbol{q}} \end{aligned}$$

where $|\dot{\boldsymbol{r}}$ is a column matrix with six zero elements. If we multiply from the left by $|\mathfrak{D}^{-1}$ we get

$$(\dot{\boldsymbol{q}} = |\mathfrak{D}^{-1}|\dot{\mathfrak{R}}$$

which permits us to express $|\boldsymbol{T}$ in terms of the time derivatives of the internal coordinates. Thus

$$\begin{aligned} 2|\boldsymbol{T} &= (\dot{\boldsymbol{q}}'\,(\dot{\boldsymbol{q}} \\ &= \dot{\mathfrak{R}}(|\mathfrak{D}^{-1})'\,|\mathfrak{D}^{-1}\dot{\mathfrak{R}} \end{aligned} \tag{5-23}$$

This follows since the transpose of a matrix composed of submatrices is obtained by interchanging rows and columns of the transposed matrices.

Upon substitution into Equation (5-23) we get

$$2|\boldsymbol{T} = (|\dot{\boldsymbol{R}}'\,|\dot{\boldsymbol{r}}') \begin{pmatrix} (\boldsymbol{Q}' \\ (\boldsymbol{Q}_0' \end{pmatrix} ((\boldsymbol{Q}\,(\boldsymbol{Q}_0) \begin{pmatrix} |\dot{\boldsymbol{R}} \\ |\dot{\boldsymbol{r}} \end{pmatrix}$$

which upon expansion yields

$$2|\boldsymbol{T} = |\dot{\boldsymbol{R}}'\,(\boldsymbol{Q}'\,(\boldsymbol{Q}\,|\dot{\boldsymbol{R}} + |\dot{\boldsymbol{R}}'\,(\boldsymbol{Q}'\,(\boldsymbol{Q}_0\,|\dot{\boldsymbol{r}} + |\dot{\boldsymbol{r}}'\,(\boldsymbol{Q}_0'\,(\boldsymbol{Q}\dot{\boldsymbol{R}} + |\dot{\boldsymbol{r}}'\,(\boldsymbol{Q}_0'\,(\boldsymbol{Q}_0\,|\dot{\boldsymbol{r}}$$

Since $\dot{\boldsymbol{r}}$ is a zero matrix,

$$2|\boldsymbol{T} = |\dot{\boldsymbol{R}}'\,(\boldsymbol{Q}'\,(\boldsymbol{Q}\,|\dot{\boldsymbol{R}}$$

We must now determine $(\boldsymbol{Q}$. The first step is to take notice of the relationships

$$\begin{aligned} |\mathfrak{D}^{-1}\,|\mathfrak{D} &= ((\boldsymbol{Q}\,(\boldsymbol{Q}_0) \begin{pmatrix} |\boldsymbol{D} \\ |\boldsymbol{D}_0 \end{pmatrix} \\ &= (\boldsymbol{Q}\,|\boldsymbol{D} + (\boldsymbol{Q}_0\,|\boldsymbol{D}_0 \\ &= |\boldsymbol{I}_{3N \times 3N} \end{aligned}$$

and

$$|\mathfrak{D}|\mathfrak{D}^{-1} = \begin{pmatrix}|\boldsymbol{D}\\ |\boldsymbol{D}_0\end{pmatrix}((\boldsymbol{Q}\,(\boldsymbol{Q}_0)$$

$$= \begin{pmatrix}|\boldsymbol{D}\,(\boldsymbol{Q} & |\boldsymbol{D}\,(\boldsymbol{Q}_0\\ |\boldsymbol{D}_0\,(\boldsymbol{Q} & |\boldsymbol{D}_0\,(\boldsymbol{Q}_0\end{pmatrix}$$

$$= |\boldsymbol{I}_{3N\times 3N}$$

Note that the matrix $|\boldsymbol{D}\,(\boldsymbol{Q}$ which is made up of the $3N - 6 \times 3N$ matrix $|\boldsymbol{D}$ and the $3N \times 3N - 6$ matrix $(\boldsymbol{Q}$ is a $(3N - 6) \times (3N - 6)$ identity matrix.

Taking the transpose of the result

$$(\boldsymbol{Q}\,|\boldsymbol{D} + (\boldsymbol{Q}_0\,|\boldsymbol{D}_0 = |\boldsymbol{I}_{3N\times 3N}$$

we have

$$|\boldsymbol{D}'\,(\boldsymbol{Q}' + |\boldsymbol{D}_0'\,(\boldsymbol{Q}_0' = |\boldsymbol{I}_{3N\times 3N}$$

Upon multiplying on the left by $|\boldsymbol{D}$ and on the right by $(\boldsymbol{Q}$ we get

$$|\boldsymbol{D}\,|\boldsymbol{D}'\,(\boldsymbol{Q}'\,(\boldsymbol{Q} + |\boldsymbol{D}\,|\boldsymbol{D}_0'\,(\boldsymbol{Q}_0'\,(\boldsymbol{Q} = |\boldsymbol{D}\,(\boldsymbol{Q} = |\boldsymbol{I}_{(3N-6)\times(3N-6)}$$

Since vibrations are orthogonal to rotations and translations, $|D$ and $|D_0$ are orthogonal, and the term $|\boldsymbol{D}\,|\boldsymbol{D}_0'\,(\boldsymbol{Q}_0'\,(\boldsymbol{Q}$ disappears. Thus

$$|\boldsymbol{D}\,|\boldsymbol{D}'\,(\boldsymbol{Q}'\,(\boldsymbol{Q} = |\boldsymbol{I}_{(3N-6)\times(3N-6)}$$

and

$$(\boldsymbol{Q}'\,(\boldsymbol{Q} = (|\boldsymbol{D}\,|\boldsymbol{D}')^{-1}$$

The matrix $(|\boldsymbol{D}\,|\boldsymbol{D}')^{-1}$ is usually designated $\boldsymbol{G}^{-1}$ where

$$G_{kl} = \sum_{i=1}^{3n} \frac{B_{ki}B_{li}}{m_i} \tag{5-24}$$

In matrix notation

$$(\boldsymbol{G} = |\boldsymbol{B}\,|\boldsymbol{M}^{-1}\,|\boldsymbol{B}' \tag{5-25}$$

We have succeeded in expressing $|\boldsymbol{T}$ and $(\boldsymbol{U}$ in terms of the internal coordinates. The expressions are

$$2\,|\boldsymbol{T} = |\dot{\boldsymbol{R}}'\,(\boldsymbol{G}^{-1}\,|\dot{\boldsymbol{R}}$$

and

$$2\,|\boldsymbol{V} = |\boldsymbol{R}'\,|\boldsymbol{F}\,|\boldsymbol{R}$$

Upon application of Lagrange's equations we expect solutions of the form

$$R_k = A_k \cos 2\,\pi v t$$

The determinant of the coefficients of the A_k is set equal to zero and there results the equation

$$\begin{vmatrix} F_{11} - G_{11}^{-1}\lambda & F_{12} - G_{12}^{-1}\lambda & \cdots \\ F_{21} - G_{21}^{-1}\lambda & \cdots & \cdots \\ \cdots & \cdots & \cdots \end{vmatrix} = 0$$

where, if $|\Lambda$ is a diagonal matrix of the solutions, λ, we get

$$\left| |\boldsymbol{F} - (\boldsymbol{G}^{-1} |\Lambda \right| = 0$$

Upon multiplication on the left by $(\boldsymbol{G}$, Equation (5-26) results.

$$\left| (\boldsymbol{G} |\boldsymbol{F} - |\Lambda \right| = 0 \qquad \textbf{(5-26)}$$

There are two ways in which this equation is used. If the frequencies of the fundamental transitions are known, then the force constants may be calculated. On the other hand, if force constants can be borrowed from similar molecules, frequencies of fundamental vibrations may be calculated.

5-12 TREATMENT OF THE H_2O MOLECULE

In this section we will calculate the force constants for the H_2O molecule. The coordinate system is shown in Figure 5.6. From Equation (5-25) we see that the $(\boldsymbol{G}$ matrix is defined as

$$(\boldsymbol{G} = |\boldsymbol{B} |\boldsymbol{M}^{-1} |\boldsymbol{B}'$$

where $|\boldsymbol{B}$ is a matrix that relates the internal displacement coordinates to Cartesian coordinates, and $|\boldsymbol{M}^{-1}$ is a matrix whose diagonal elements are μ_i, the reciprocal of the mass of the atom being displaced along the ith coordinate. As defined then

$$|\boldsymbol{R} = |\boldsymbol{B} |\mathbf{X}$$

and we now wish to construct the matrix $|\boldsymbol{B}$. Using the coordinate system shown in Figure 5.6 it may readily be seen that the internal coordinates Δr_a, Δr_b, and $\Delta(2\alpha)$ are related to the Cartesian coordinates in the following way:

$$\begin{pmatrix} \Delta r_a \\ \Delta r_b \\ \Delta(2\alpha) \end{pmatrix} = \begin{pmatrix} -\sin\alpha & -\cos\alpha & 0 & 0 & \sin\alpha & \cos\alpha \\ 0 & 0 & \sin\alpha & -\cos\alpha & -\sin\alpha & \cos\alpha \\ -\frac{1}{r}\cos\alpha & \frac{1}{r}\sin\alpha & \frac{1}{r}\cos\alpha & \frac{1}{r}\sin\alpha & 0 & -\frac{2}{r}\sin\alpha \end{pmatrix} \begin{pmatrix} \Delta y_a \\ \Delta z_a \\ \Delta y_b \\ \Delta z_b \\ \Delta y_c \\ \Delta z_c \end{pmatrix}$$

If we let $s = \sin\alpha$ and $c = \cos\alpha$, then $(G$ becomes

$$(G = \begin{pmatrix} -s & -c & 0 & 0 & s & c \\ 0 & 0 & s & -c & -s & c \\ -\frac{c}{r} & \frac{s}{r} & \frac{c}{r} & \frac{s}{r} & 0 & -\frac{2s}{r} \end{pmatrix}$$

$$\times \begin{pmatrix} \mu_a & 0 & 0 & 0 & 0 & 0 \\ 0 & \mu_a & 0 & 0 & 0 & 0 \\ 0 & 0 & \mu_b & 0 & 0 & 0 \\ 0 & 0 & 0 & \mu_b & 0 & 0 \\ 0 & 0 & 0 & 0 & \mu_c & 0 \\ 0 & 0 & 0 & 0 & 0 & \mu_c \end{pmatrix} \begin{pmatrix} -s & 0 & -\frac{c}{r} \\ -c & 0 & \frac{s}{r} \\ 0 & s & \frac{c}{r} \\ 0 & -c & \frac{s}{r} \\ s & -s & 0 \\ c & c & -\frac{2s}{r} \end{pmatrix}$$

$$= \begin{pmatrix} \mu_a + \mu_c & \mu_c \cos 2\alpha & -\frac{\mu_c}{r}\sin 2\alpha \\ \mu_c \cos 2\alpha & \mu_b + \mu_c & -\frac{\mu_c}{r}\sin 2\alpha \\ -\frac{\mu_c}{r}\sin 2\alpha & -\frac{\mu_c}{r}\sin 2\alpha & \frac{\mu_a}{r^2} + \frac{\mu_b}{r^2} + \frac{2\mu_c}{r^2}(1 - \cos 2\alpha) \end{pmatrix}$$

The $(G$ matrix may now be symmetry factored by use of the $|U$ matrices which were introduced in Section 5-6.

For the H_2O molecule, or any XY_2 molecule belonging to point group C_{2v}, the symmetry coordinates are

$$s_1(A_1) = \frac{1}{\sqrt{2}}(\Delta r_a + \Delta r_b)$$

$$s_2(A_1) = \Delta(2\alpha)$$

$$s_3(B_2) = \frac{1}{\sqrt{2}}(\Delta r_a - \Delta r_b)$$

or in matrix form

$$\begin{pmatrix} s_1 \\ s_2 \\ s_3 \end{pmatrix} = \begin{pmatrix} \frac{1}{\sqrt{2}} & \frac{1}{\sqrt{2}} & 0 \\ 0 & 0 & 1 \\ \frac{1}{\sqrt{2}} & -\frac{1}{\sqrt{2}} & 0 \end{pmatrix} \begin{pmatrix} \Delta r_a \\ \Delta r_b \\ \Delta(2\alpha) \end{pmatrix}$$

For ease of manipulation, ($\mathbf{G}$ may be represented as

$$(\mathbf{G} = \begin{pmatrix} A & B & C \\ B & A & C \\ C & C & D \end{pmatrix}$$

where

$$A = \mu_a + \mu_c = \mu_b + \mu_c$$

$$B = \mu_c \cos 2\alpha$$

$$C = -\frac{\mu_c}{r} \sin 2\alpha$$

$$D = \frac{\mu_a}{r} + \frac{\mu_b}{r} + \frac{2\mu_c}{r}(1 - \cos 2\alpha)$$

Carrying out the transformation we get

$$\begin{pmatrix} \frac{1}{\sqrt{2}} & \frac{1}{\sqrt{2}} & 0 \\ 0 & 0 & 1 \\ \frac{1}{\sqrt{2}} & -\frac{1}{\sqrt{2}} & 0 \end{pmatrix} \begin{pmatrix} A & B & C \\ B & A & C \\ C & C & D \end{pmatrix} \begin{pmatrix} \frac{1}{\sqrt{2}} & 0 & \frac{1}{\sqrt{2}} \\ \frac{1}{\sqrt{2}} & 0 & -\frac{1}{\sqrt{2}} \\ 0 & 1 & 0 \end{pmatrix}$$

$$= \begin{pmatrix} A + B & \sqrt{2}C & 0 \\ \sqrt{2}C & D & 0 \\ 0 & 0 & A - B \end{pmatrix}$$

where the (**G** matrix has been block diagonalized into a 2×2 for the A_1 normal modes and a 1×1 matrix for the B_1 normal mode.

The **F** matrix may be symmetry factored in the same manner to yield

$$\begin{pmatrix} \frac{1}{\sqrt{2}} & \frac{1}{\sqrt{2}} & 0 \\ 0 & 0 & 1 \\ \frac{1}{\sqrt{2}} & -\frac{1}{\sqrt{2}} & 0 \end{pmatrix} \begin{pmatrix} f_r & f_{rr} & rf_{r\alpha} \\ f_{rr} & f_r & rf_{r\alpha} \\ rf_{r\alpha} & rf_{r\alpha} & r^2 f_\alpha \end{pmatrix} \begin{pmatrix} \frac{1}{\sqrt{2}} & 0 & \frac{1}{\sqrt{2}} \\ \frac{1}{\sqrt{2}} & 0 & -\frac{1}{\sqrt{2}} \\ 0 & 1 & 0 \end{pmatrix}$$

$$= \begin{pmatrix} f_r + f_{rr} & \sqrt{2}\, rf_{r\alpha} & 0 \\ \sqrt{2}\, rf_{r\alpha} & r^2 f_\alpha & 0 \\ 0 & 0 & f_r - f_{rr} \end{pmatrix}$$

We may now substitute the |**F** and (**G** matrices into the equation $|(\boldsymbol{G}|\boldsymbol{F} - |\boldsymbol{\Lambda}| = 0$, and solve for the frequencies of the vibrations, or in this case fit the observed frequencies to force constants. First consider the B_2 mode. Upon substitution there results the linear equation

$$(f_r - f_{rr})(A - B) - \lambda = 0$$

from which

$$f_r - f_{rr} = \frac{\lambda}{\mu_b + \mu_c\, 2 \sin^2 \alpha} \qquad \textbf{(5-27)}$$

For the A_1 normal modes the equation becomes

$$\begin{vmatrix} G_{11}F_{11} + G_{12}F_{21} - \lambda & G_{11}F_{12} + G_{12}F_{22} \\ G_{21}F_{11} + G_{22}F_{21} & G_{21}F_{12} + G_{22}F_{22} - \lambda \end{vmatrix} = 0$$

Expansion of the determinant yields

$$\lambda^2 - (G_{11}F_{11} + G_{22}F_{22} + 2G_{12}F_{12})\lambda + (G_{11}G_{22} - G_{12}^2)(F_{11}F_{22} - F_{12}^2) = 0 \qquad \textbf{(5-28)}$$

The observed frequencies are $\nu_1 = 3825$, $\nu_2 = 1654$, and $\nu_3 = 3936\ \text{cm}^{-1}$. With three pieces of data, the four force constants in the potential energy expression can not be determined uniquely. Of the four force constants let us assume that f_{rr} is the smallest and can be ignored. Thus, from the B_2 mode we can calculate f_r as follows from Equation (5-27):

$$f_r = \frac{4\pi^2 c^2 \nu^2}{\mu_1 + \mu_3(1 - \cos 2\alpha)}$$
$$= 8.5 \times 10^5\ \text{dyne cm}^{-1}$$

With this value for f_r, the two A_1 energies may be used in Equation (5-28) to give the other two force constants. The results are

$$f_{r\alpha} = .25 \times 10^5 \text{ dyne cm}^{-1}$$
$$f_\alpha = .77 \times 10^5 \text{ dyne cm}^{-1}$$

It is apparent that the computational difficulties will increase as the molecules become more complex. For this reason the use of digital computers in solution of vibrational problems is very common. Computer programs for normal coordinate analyses can be obtained from the Quantum Chemistry Program Exchange Center at the University of Indiana.

5-13 GENERAL FORMULAS FOR G MATRIX ELEMENTS

The most tedious task in the normal coordinate treatment is the construction of the (**G** matrix. Contemplation of the derivation of the (**G** matrix suggests that general formulas for the (**G** matrix elements are possible since, for example, the matrix element for a bond stretching vibration is not dependent on the coordinant system used. Such general formulas have been derived and are discussed in detail in specialized books. Here we present a brief introduction which demonstrates the utility of these formulas.

A number of symbols must first be introduced. Atoms common to the two internal coordinates are put on a horizontal line and are noted as such by a set of two concentric circles. For example, consider the (**G** matrix element for a bond stretching motion. The diagram is

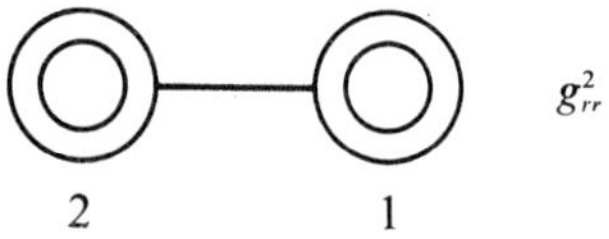

The atoms are numbered in a uniform way. Common atoms are numbered first, and they are numbered from right to left. The matrix element is designated by the letter g which is given a double subscript to denote the type of internal coordinates involved. A post superscript specifies the number of atoms common to the two internal coordinates. Thus, for a bond stretching coordinate the designation is g_{rr}^2. For the interaction element connecting two bond stretching coordinates with a common atom the appropriate diagram and symbol becomes

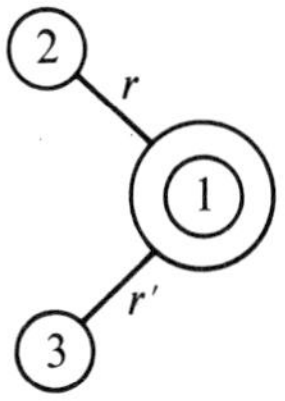

g^1_{rr}

It is important to recognize that the actual molecular geometry is ignored when diagramming the (***G*** matrix element.

Now consider the matrix elements for the angle deformations. For the diagonal matrix element the diagram and symbol are

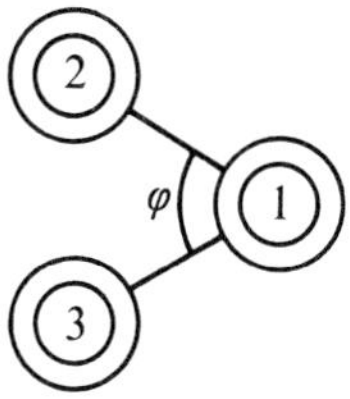

$g^3_{\varphi\varphi}$

The meaning of symbol $g^3_{\phi\phi}$ should be clear. These are three common atoms, and the subscript indicates an angle deformation. Interaction elements present a new problem. We can see that three possibilities arise for the element $g^1_{\phi\phi}$. The common atom could be (1) terminal for both internal coordinates, (2) terminal for one coordinate and central for the second coordinate, or (3) central to both coordinates. In order to differentiate these interaction elements, the noncommon atoms are put on 45° diagonals in such a way that the maximum number of bonded atoms, m, lies along the upper left hand diagonal, and the next largest number, n, lies along the lower left hand diagonal, and the symbol is supplemented by the column matrix $\begin{pmatrix} m \\ n \end{pmatrix}$. In the three cases mentioned above for $g^1_{\phi\phi}$, we have

CASE 1

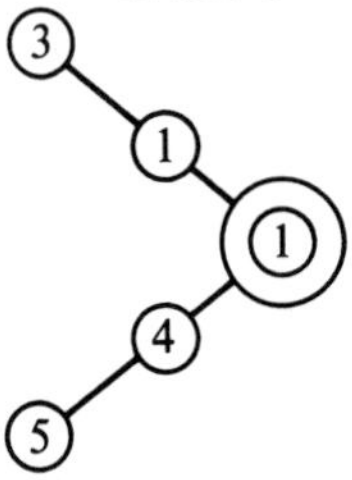

$g^{1}_{\varphi\varphi}\binom{2}{2}$

CASE 2

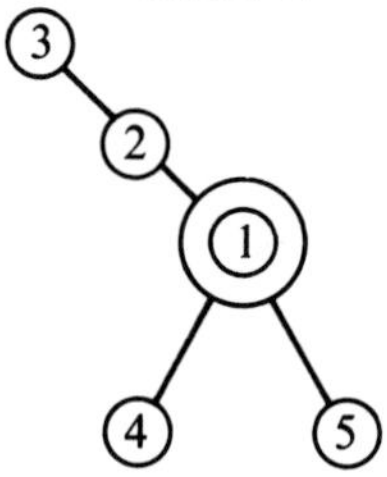

$g^{1}_{\varphi\varphi}\binom{2}{1}$

CASE 3

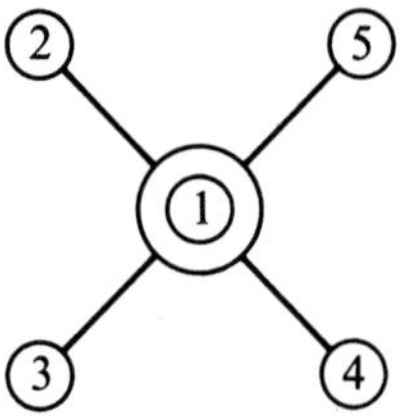

$g^{1}_{\varphi\varphi}\binom{1}{1}$

Some additional symbols and diagrams which should be readily identified are

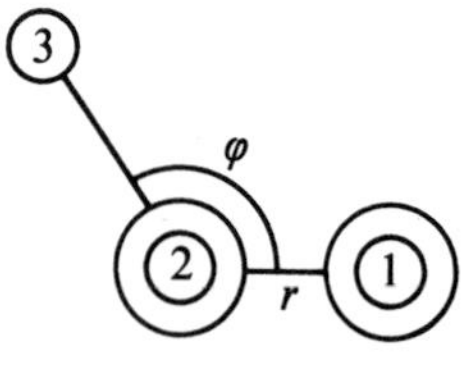

$g^{2}_{r\varphi}$

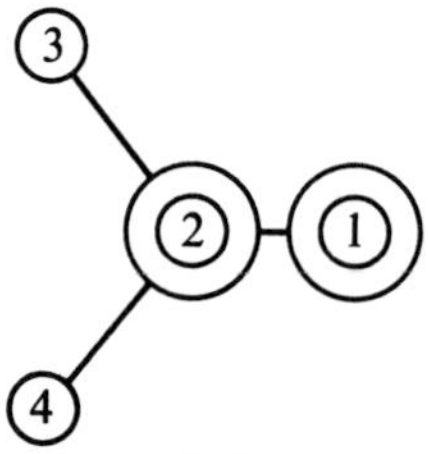

$g^{2}_{\varphi\varphi'}\binom{1}{1}$

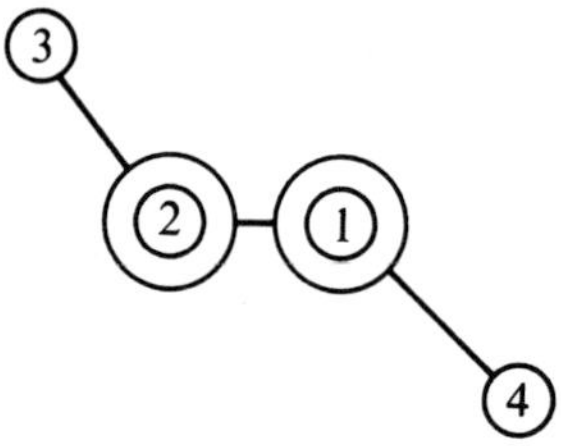

$g^{2}_{\varphi\varphi}\binom{1}{0}$

Inspection of the expression $G^{2}_{QQ'}\begin{pmatrix}1\\1\end{pmatrix}$, which is given in Table 5-2, reveals that the solid angle $\psi_{\alpha\beta\gamma}$ must be defined. The definition is as follows:

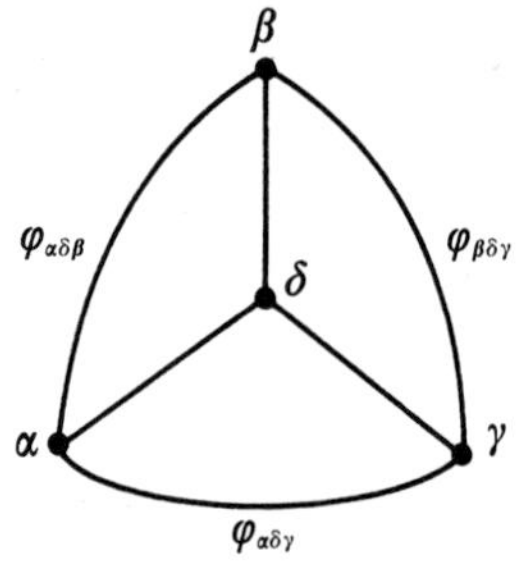

$$\cos\psi_{\alpha\beta\gamma} = \frac{\cos\varphi_{\alpha\delta\gamma} - (\cos\varphi_{\alpha\delta\beta})(\cos\varphi_{\beta\delta\gamma})}{(\sin\varphi_{\alpha\delta\beta})(\sin\varphi_{\beta\delta\gamma})}$$

Table 5.2
*G Matrix Elements**

$$g_{rr}^2 = \mu_1 + \mu_2 \qquad g_{rr}^1 = \mu_1 c\phi$$

$$g_{\phi\phi}^3 = \rho_{12}^2\mu_1 + \rho_{23}^2\mu_3 + (\rho_{12}^2 + \rho_{23}^2 - 2\rho_{12}\rho_{23}c\phi)\mu_2$$

$$g_{rr}^4 = \frac{\rho_{12}^2}{s^2\phi_2}\mu_1 + \frac{\rho_{34}^2}{s^2\phi_3}\mu_4 + \left[\left(\frac{\rho_{12} - \rho_{23}c\phi_2}{s\phi_2}\right)^2 + \frac{\rho_{23}^2}{s^2\phi_3}c^2\phi_3 - \frac{2(\rho_{12} - \rho_{23}c\phi_2)}{s\phi_2}\frac{\rho_{23}c\phi_3}{s\phi_3}c\chi\right]\mu_2 + \left[\left(\frac{\rho_{34} - \rho_{23}c\phi_3}{s\phi_3}\right)^2 + \rho_{23}^2\frac{c^2\phi_2}{s^2\phi_2} - \frac{2(\rho_{34} - \rho_{23}c\phi_3)}{s\phi_3}\frac{\rho_{23}c\phi_2}{s\phi_2}c\chi\right]\mu_3$$

$$g_{r\phi}^2 = -\rho_{23}s\phi\mu_2$$

$$g_{r\phi}^1\binom{1}{2} = \rho_{13}s\phi_1 c\chi\mu_1,$$

$$g_{r\phi}^1\binom{1}{1} = -(\rho_{13}s\phi_{213}c\psi_{234} + \rho_{14}s\phi_{214}c\psi_{243})\mu_1$$

$$g_{\phi\phi}^2\binom{1}{1} = \rho_{12}^2 c\psi_{314}\mu_1 + [(\rho_{12} - \rho_{23}c\phi_{123} - \rho_{24}c\phi_{124})\rho_{12}c\psi_{314} + (s\phi_{123}s\phi_{124}s^2\psi_{314} + c\phi_{324}c\psi_{314})\rho_{23}\rho_{24}]\mu_2$$

$$g_{\phi\phi}^2\binom{1}{0} = -\rho_{12}c\chi[(\rho_{12} - \rho_{14}c\phi_1)\mu_1 + (\rho_{12} - \rho_{23}c\phi_2)\mu_2]$$

$$g_{\phi\phi}^1\binom{2}{2} = -(s\chi_{25}s\chi_{34} + c\chi_{25}c\chi_{34}c\phi_1)\rho_{12}\rho_{14}\mu_1$$

$$g_{\phi\phi}^1\binom{2}{1} = [(s\phi_{214}c\phi_{415}c\chi_{34} - s\phi_{215}s\chi_{35})\rho_{14} + (s\phi_{215}c\phi_{415}c\chi_{35} - s\phi_{214}c\chi_{34})\rho_{15}]\frac{\rho_{12}\mu_1}{s\phi_{415}}$$

*Note: In this table the following abbreviations appear: s, sin; c, cos; ρ_{ij} is the inverse bond distance; χ is the dihedral angle. Reprinted with permission from D. Steele, *Theory of Vibrational Spectroscopy*, Philadelphia: W. B. Saunders Co., 1971.

The dihedral angle χ is defined as that angle between the planes described by the atom sets 1, 2, 3 and 2, 3, 4, where atoms 1, 2, 3, 4 are bonded in sequence. This angle is always selected so that $0 \leqslant \chi \leqslant \pi$. When viewing along the bond 2-3 from atom 2, the projection of 3-4 lies clockwise to 2-1.

We conclude this section by selecting the G matrix elements for the trigonal pyramidal AB_3 molecule. The results are shown in Figure 5.8.

	Δr_{12}	Δr_{13}	Δr_{14}	$\Delta \alpha_{213}$	$\Delta \alpha_{314}$	$\Delta \alpha_{214}$
Δr_{12}	g_{rr}^{2}	g_{rr}^{1}	g_{rr}^{1}	$g_{r\phi}^{2}$	$g_{r\phi}^{1}\begin{pmatrix}1\\1\end{pmatrix}$	$g_{r\phi}^{2}$
Δr_{13}		g_{rr}^{2}	g_{rr}^{1}	$g_{r\phi}^{2}$	$g_{r\phi}^{2}$	$g_{r\phi}^{1}\begin{pmatrix}1\\1\end{pmatrix}$
Δr_{14}			g_{rr}^{2}	$g_{r\phi}^{1}\begin{pmatrix}1\\1\end{pmatrix}$	$g_{r\phi}^{2}$	$g_{r\phi}^{2}$
$\Delta \alpha_{213}$				$g_{\phi\phi}^{3}$	$g_{\phi\phi}^{2}\begin{pmatrix}1\\1\end{pmatrix}$	$g_{\phi\phi}^{2}\begin{pmatrix}1\\1\end{pmatrix}$
$\Delta \alpha_{314}$					$g_{\phi\phi}^{3}$	$g_{\phi\phi}^{2}\begin{pmatrix}1\\1\end{pmatrix}$
$\Delta \alpha_{214}$						$g_{\phi\phi}^{3}$

Figure 5.8
G matrix for the trigonal pyramidal molecule AB_3 (point group C_{3v})

5-14 FUNCTIONAL GROUP FREQUENCIES

Since infrared spectra are frequently used for structural and bonding characterizations, it is absolutely necessary that the practicing chemist have a familiarity with the frequencies of vibrations associated with certain easily identifiable molecular fragments. Although the energies of the vibrations of a molecular fragment or functional groups are dependent on the other atoms to which the group is bonded, it is frequently found that a particular functional group will give rise to an absorption band in a limited region of the spectrum irrespective of the molecular environment.

For efficient use of vibrational spectral data, charts showing commonly occurring functional group frequencies should be readily available, and the most important ones of these committed to memory. Valuable charts showing the ranges of group frequencies may be found in various monographs devoted to vibrational spectroscopy. It is not our intention to present the vast amount of data which would be necessary for the compilation of such charts, and we emphasize that proper and meaningful use of functional group frequency ranges requires careful study of the spectral data of series of molecules which contain these groups. There is only one justifiable approach to use of these charts; the ranges of frequencies should be verified or compiled by the user. There is hardly a pressing need for a carbon chemist to know the intricate details of metal-ligand vibrations, although he may wish to know in what range of the spectrum such absorption bands may be found. We propose that the reader design and construct his own correlation charts by reference to the original literature or to appropriate monographs.

5-15 PROBLEMS

1. The Raman spectrum of BrF_5 exhibits nine absorptions which may be attributed to fundamental bands. Are the observations consistent with a trigonal bipyramidal or tetragonal pyramidal structure?
2. The Raman spectrum of IF_7 exhibits five absorption bands, as does the infrared spectrum. Are these observations consistent with a pentagonal bipyramidal structure for IF_7?
3. Two likely structures for N_2F_4 are

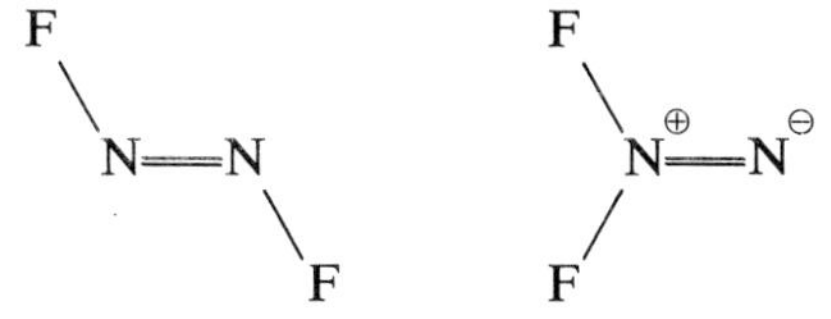

C_{2h} C_{2v}

There are three fundamental bands in the IR, and three fundamental bands in the Raman spectrum. The IR and Raman bands do not coincide in energy. On the basis of this information which is the more likely structure?

4. A terminal carbonyl group absorbs at 2000–2100 cm^{-1} while a bridging carbonyl group absorbs at 1800–1900 cm^{-1}. In view of the fact that the infrared spectrum of $Fe_2(CO)_9$ exhibits bands at 1828, 2034, and 2080 cm^{-1}, deduce a structure for the molecule.
5. The uncoordinated perchlorate ion exhibits a strong absorption band at 1100 cm^{-1} and a very weak absorption at 930 cm^{-1}. Upon coordination (as a unidentate ligand) the band at 1100 cm^{-1} splits into two bands, and the band at 930 cm^{-1} gains considerable intensity. Account for these observations on the basis of symmetry arguments.
6. Assume that the force constants obtained for the H_2O molecule in Section 5-12 are valid for the D_2O molecule and predict the energies of the fundamental modes of vibration for D_2O. Do your results agree very well with the experimental values for D_2O?

7. The C–N stretching vibration in the cyanide ion gives rise to a relatively sharp but somewhat weak band in the region 2050–2250 cm^{-1}. In ionic compounds such as NaCN the band occurs at 2080 cm^{-1}, but in covalently bonded systems the energy depends dramatically on the atom bonded to carbon. For example, in $P(CN)_3$, the stretching vibration is at 2204 cm^{-1}. Account for this observation.
8. In view of the fact that ClF_3 exhibits six infrared bands, predict a structure.
9. The infrared and Raman spectra of $^{10}BF_3$ exhibit the following bands:

IR
482(*s*) 720(*s*) ··· 1497(*vs*) 1928(*w*) 2058(*w*) 2250(*w*) 3008(*w*) 3260(*w*)

Raman
482(*m*) ··· 888(*s*) ————————————
(not studied)

Assign these bands according to the proper symmetry species and label them with the appropriate ν_n. Check the selection rules to be sure the combinations and overtones are allowed.

10. Infrared and Raman fundamental modes of vibration for the series of deuteromethanes are tabulated in Table 5.3. Assign these bands appropriate ν_n symbols and symmetry labels. Trace the splitting and coalescence of the fundamental modes through the series by means of a correlation chart.

Table 5.3
Vibrational Spectral Data for CH_nD_{4-n}

CH_4	CH_3D	CH_2D_2	CHD_3	CD_4
3020 I, R	3030 I	3020 I	2992 I	2258 I, R
2914 R	2982 I	2974 R	2267 I, R	2085 R
1526 I, R	2205 I, R	2255 I	2141 R	1054 R
1306 I	1477 I, R	2139 R	1299 I, R	996 I
	1306 I, R	1450 R	1046 I, R	
	1156 I, R	1286 R	982 I, R	
		1235 I		
		1090 R		
		1034 I, R		

5-16 REFERENCES

Excellent textbooks and reference books on infrared spectroscopy and its application to inorganic chemistry are available. Some particularly useful ones are:

G. M. Barrow, *Molecular Spectroscopy.* New York: McGraw Hill Book Co., 1962.

G. Herzberg, *Infrared and Raman Spectra.* New York: D. Van Nostrand., 1945.

K. Nakamoto, *Infrared Spectra of Inorganic and Coordination Compounds.* 2d ed. New York: John Wiley. 1970.

E.B. Wilson, J.C. Decius, and P.C. Cross, *Molecular Vibrations,* New York: McGraw-Hill, 1955.

6

CRYSTAL FIELD THEORY AND LIGAND FIELD THEORY

The electronic spectra of transition metal complexes are very valuable for characterization purposes since they can be used to study many important chemical and physical effects. Electronic spectra are conveniently systematized by the closely related ligand field and crystal field theories, and symmetry notions are very helpful in explaining such facts as the number of bands in the spectra, the relative intensities of spectral bands, and occasionally the relative energies of these bands. In this chapter we will develop certain aspects of ligand field and crystal field theories that prove useful in subsequent studies of descriptive chemistry.

6-1 MODELS OF THE ATOM

Crystal field and ligand field theories utilize the results of atomic theory extensively. Specifically in the calculations and manipulations of the theory for transition metal ions, the angular portions of the atomic wave functions are used, but the radial parts of the wave functions are frequently taken as parameters. That is, in the general form of the hydrogen-like wave function which is given in Equation (6-1), the $\Theta(\vartheta)$ and $\Phi(\varphi)$ parts of the function are retained but $R(r)$ is treated as a parameter. This concept is treated in detail in Section 6-5. Some useful angular wave functions are collected in Table 4.1. It is now necessary to develop some additional concepts concerning many-electron atoms to supplement the material developed in Chapter 4.

$$\psi_{n,l,m} = \Theta(\vartheta)\Phi(\varphi)\,R(r) \tag{6-1}$$

The Schrödinger wave equation for a two-electron atom can be written explicitly as

$$-\frac{\hbar^2}{2m}\nabla_1^2\psi - \frac{\hbar^2}{2m}\nabla_2^2\psi - \left(E + \frac{Ze^2}{r_1} + \frac{Ze^2}{r_2} - \frac{e^2}{r_{12}}\right)\psi = 0 \tag{6-2}$$

where the subscripts specify the coordinates of the two electrons and the nucleus is taken to be at the origin of the coordinate system; six coordinates are required to describe the positions of the two electrons. Extension of the theory to many electrons is easily done, where additional kinetic energy, nuclear attraction, and interelectronic repulsion terms appear. However, the Schrödinger equation for even two electrons can not be solved exactly because of the interelectronic repulsion term. It is necessary to employ specialized techniques for finding approximate solutions of the Schrödinger equation for these more complex problems. The three most widely used techniques are

1. the perturbation method
2. the variation method
3. the Hartree-Fock self-consistent-field method

We shall consider only the perturbation method here in a treatment of the two-electron problem that arises in the case of the helium atom.

If the interelectronic repulsion term e^2/r_{12} in Equation (6-2) is considered to be a small energy term, then it may be treated as a perturbation. In the usual manner this term is first ignored and it can be seen that the Schrödinger equation reduces to

$$-\frac{\hbar^2}{2m}\nabla_1^2\psi - \frac{\hbar^2}{2m}\nabla_2^2\psi - \left(\frac{Ze^2}{r_1} + \frac{Ze^2}{r_2}\right)\psi = E\psi \tag{6-3}$$

Next we define the two-electron wave function to be

$$\psi = \psi_1(r_1, \vartheta_1, \varphi_1)\psi_2(r_2, \vartheta_2, \varphi_2)$$

and we find that Equation (6-3) becomes

$$\psi_2\left[-\frac{\hbar^2}{2m}\nabla_1^2 - \frac{Ze^2}{r_1}\right]\psi_1 + \psi_1\left[-\frac{\hbar^2}{2m}\nabla_2^2 - \frac{Ze^2}{r_2}\right]\psi_2 = E\psi_1\psi_2 \tag{6-4}$$

If we say that $E = E_1 + E_2$, then upon dividing Equation (6-4) through by $\psi_1 \psi_2$, we get

$$\frac{1}{\psi_1}\left[-\frac{\hbar^2}{2m}\nabla_1^2 - \frac{Ze^2}{r_1}\right]\psi_1 + \frac{1}{\psi_2}\left[-\frac{\hbar^2}{2m}\nabla_2^2 - \frac{Ze^2}{r_2}\right]\psi_2 = E_1 + E_2$$

which is related to the two equations

$$\left[-\frac{\hbar^2}{2m}\nabla_1^2 - \frac{Ze^2}{r_1}\right]\psi_1 = E_1\psi_1 \quad \text{and} \quad \left[-\frac{\hbar^2}{2m}\nabla_2^2 - \frac{Ze^2}{r_2}\right]\psi_2 = E_2\psi_2$$

These are immediately seen to be Schrödinger equations for one electron, so

$$E_1 = -\frac{m_2 Z^2 e^4}{\hbar^2 n_1^2} \quad \text{and} \quad E_2 = -\frac{m_2 Z^2 e^4}{\hbar^2 n_2^2}$$

Thus, for the ground state where $n_1 = n_2 = 1$, we may write for the wave function

$$\psi = \psi_1(r_1, \vartheta_1, \varphi_1)\psi_2(r_2, \vartheta_2, \varphi_2) = \psi(1s)_1(1s)_2 \tag{6-5}$$

We have found that the energy of the helium atom may be approximated in the absence of interelectronic repulsion effects to be two times the energy of the hydrogen-like atom with nuclear charge Z. In perturbation theory terminology this energy is the zero-order energy and the function in Equation (6-5) is the zero-order function. To calculate the first-order correction to the energy, the integral in Equation (6-6) would have to be solved.

$$E^1 = \iint \psi(1s)_1^* \psi(1s)_2^* (e^2/r_{12})\psi(1s)_1 \psi(1s)_2 \, d\tau_1 \, d\tau_2 \tag{6-6}$$

This has been done, and as shown in Eyring, Walter, and Kimball* the integral is equal to $(5/4)ZE_H$. The energy of the helium atom is then

$$E = -2Z^2E_H + (5/4)ZE_H = -(11/2)E_H = -74.8 \; eV$$

The experimental value of 78.98 eV may be more closely approximated by the variation method, but we will leave the subject here since we have seen that approximate wave functions for many-electron atoms may be built up from one-electron wave functions. The zero-order wave functions are the products of the appropriate hydrogen-like wave functions.

In many-electron atoms for a given principal quantum shell the energies of the orbitals depend on the l quantum number. For example, p electrons

*H. Eyring, J. Walter, and G. E. Kimball, *Quantum Chemistry* (New York: John Wiley and Sons, 1944).

are, on the average, farther from the nucleus than *s* electrons. The latter screen the nuclear charge experienced by *p* electrons and consequently *p* orbitals have higher energies. The relative energies of a few low-lying orbitals are shown in Figure 6.1. This ordering does not hold for all values of *Z* (the atomic number), but will suffice for the purposes of our discussions.

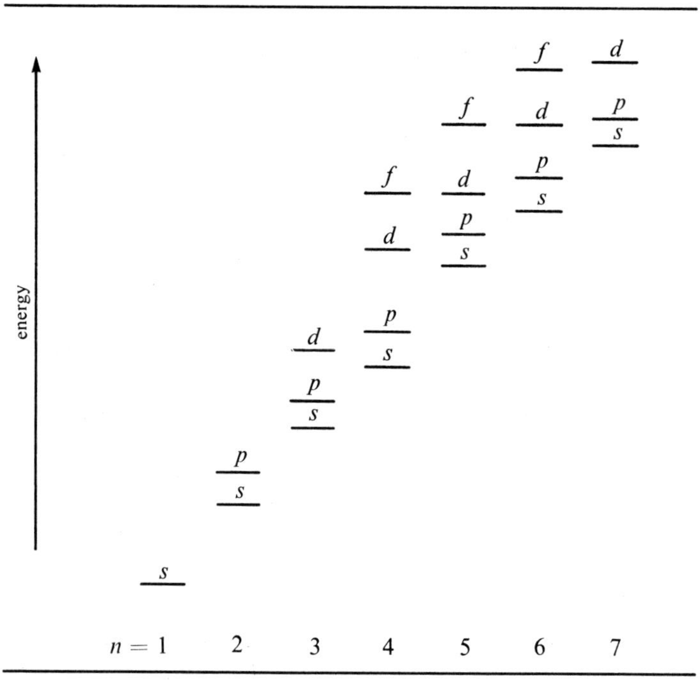

Figure 6.1
Relative Energies of a Few Low Lying Orbitals in Neutral Atoms

Up to now we have made no mention of electron spin. In the complete relativistic theory of the atom, spin quantum numbers $m_s = \pm\frac{1}{2}$ arise, and there are now four quantum numbers (n, l, m, and m_s) for the atomic wave functions. Only two electrons can be accommodated in a hydrogen-like *s* orbital, since according to the *Pauli Principle*, no two electrons can have the same four quantum numbers. Following this reasoning six electrons can be placed in *p* subshells, 10 electrons in *d* subshells, 14 electrons in *f* subshells, etc.

To describe the electronic structure of an atom, the proper number of electrons are placed in the hydrogen-like orbitals according to the *aufbau principle*. That is, electrons are placed in the lowest lying orbitals first. The arrangement of electrons in the available orbitals is called the *electronic configuration.*

A useful mnemonic for constructing electronic configurations is shown in Figure 6.2. The reader should be aware, however, that this mnemonic

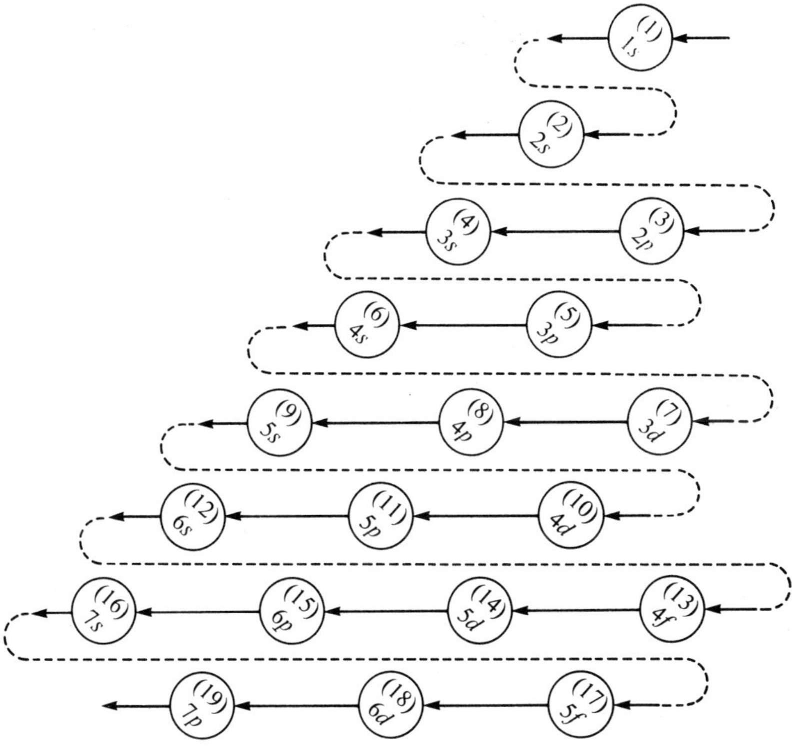

Figure 6.2
Mnemonic for Aufbau Principle

device is not infallible. Reference to tables of electronic configurations will quickly reveal deviations from the simple model.

6-2 RUSSELL-SAUNDERS COUPLING

In the previous section we have seen that an approximate wave function for many-electron atoms can be made up from one-electron functions. It was also pointed out that it is convenient to describe the electronic structure of these many-electron atoms in terms of configurations. For example, the configuration of the carbon atom in its ground state is $(1s)^2\,(2s)^2\,(2p)^2$. This configuration can give rise to various energy states depending on the orbital angular momenta and spin angular momenta of the two electrons in the unfilled p-orbital shell. The electrons in the filled orbital shells, $1s$ and $2s$, will have no effect since they contribute $M_L = 0$ and $M_S = 0$. In general, only electrons in unfilled orbital shells determine the various energy states that can result from a coupling of their respective spin-angular and orbital-angular momenta.

For the lighter atoms the spin-angular and orbital-angular momenta couple according to the Russell-Saunders scheme. In this coupling scheme, all the spin-angular momenta add vectorially, and all the orbital-angular momenta add vectorially subject to quantum restrictions.

It is common practice to use capital letters S, L, M_L, M_S to designate properties of many-electron atoms and lower case letters for individual electrons.

In order to determine systematically the number of states which can arise from a specific electronic configuration such as the carbon $(1s)^2\,(2s)^2\,(2p)^2$, we will construct a chart of all the possible combinations of the m_s's and m_l's of the two $2p$ electrons. We will write these combinations as (m_{l_1}, m_{l_2}) with a $+$ or $-$ sign on top of the m_l values to indicate the m_s quantum number. Thus, if both our electrons had $m_l = 1$, and electron 1 had $m_s = +\frac{1}{2}$ and electron 2 had $m_s = -\frac{1}{2}$, then our symbol would be $(\overset{+}{1}, \overset{-}{1})$. It might be noted that the symbol $(\overset{-}{1}, \overset{+}{1})$ is not different from $(\overset{+}{1}, \overset{-}{1})$, because the two electrons are indistinguishable. If electron 1 had $m_l = 1$, $m_s = +\frac{1}{2}$, and electron 2 had $m_l = 0$, $m_s = -\frac{1}{2}$, we would write $(\overset{+}{1}, \overset{-}{0})$. This latter symbol is different from $(\overset{-}{1}, \overset{+}{0})$, but not different from $(\overset{-}{0}, \overset{+}{1})$.

The chart is given in Figure 6.3. Since we have two p electrons, then the maximum L that we can have is $L = l_1 + l_2 = 2$ and the maximum $S = s_1 + s_2 = 1$. These give rise to the quantum numbers $M_L = 2, 1, 0, -1, -2$ and $M_s = 1, 0, -1$ which have been used to classify the symbols in the chart. The symbols, which are quantum numbers for wave functions

M_L \ M_S	*1*	*0*	*−1*
2		$(\overset{+}{1}, \overset{-}{1})$	
1	$(\overset{+}{1}, \overset{+}{0})$	$(\overset{+}{1}, \overset{-}{0})(\overset{-}{1}, \overset{+}{0})$	$(\overset{-}{1}, \overset{-}{0})$
0	$(\overset{+}{1}, -\overset{+}{1})$	$(\overset{+}{1}, -\overset{-}{1})(\overset{-}{1}, -\overset{+}{1})$ $(\overset{+}{0}, \overset{-}{0})$	$(\overset{-}{1}, -\overset{-}{1})$
−1	$(-\overset{+}{1}, \overset{+}{0})$	$(-\overset{+}{1}, \overset{-}{0})(-\overset{-}{1}, \overset{+}{0})$	$(-\overset{-}{1}, \overset{-}{0})$
−2		$(-\overset{+}{1}, -\overset{-}{1})$	

Figure 6.3
Microstate Chart for p^2 Configuration

and are called microstates, are then written in the appropriate boxes. We do not have $(\overset{+}{1}, \overset{+}{1})$ or $(\overset{-}{1}, \overset{-}{1})$ since these violate the Pauli principle. Both electrons would have all four quantum numbers the same, and that is not permitted.

Each atomic state with a given L and S encompasses a complete set of microstates from $+L$ to $-L$, i.e. the possible M_L values, and $+S$ to $-S$, the M_s values. Thus, the microstate in the left hand top corner of our chart, $(\overset{+}{1}, \overset{+}{0})$, may be considered to be the parent microstate of a set arising from a state with $L = 1$ and $S = 1$. Therefore, we must remove 9 microstates from the chart in order to account for all the possible m_l and m_s values of this state. The nine microstates are

$$\begin{array}{ccc} (\overset{+}{1}, \ \overset{+}{0}) & (\overset{+}{1}, \ \overset{-}{0}) & (\overset{-}{1}, \ \overset{-}{0}) \\ (\overset{+}{1}, -\overset{+}{1}) & (\overset{+}{1}, -\overset{-}{1}) & (\overset{-}{1}, -\overset{-}{1}) \\ (-\overset{+}{1}, \ \overset{+}{0}) & (-\overset{+}{1}, \ \overset{-}{0}) & (-\overset{-}{1}, \ \overset{-}{0}) \end{array}$$

where we have arbitrarily taken only one microstate from those boxes in the chart which contain more than one microstate.

We again go to the top left (middle) of the chart and choose the microstate $(\overset{+}{1}, \overset{-}{1})$. This is the parent microstate for a state with $L = 2$ and $S = 0$. Here we will remove five microstates from the chart accounting for the M_L values 2, 1, 0, −1, −2. (M_s can only be 0). Finally, we see that there is only one microstate left, $(\overset{+}{0}, \overset{-}{0})$. This describes in itself a state with $L = 0$ and $S = 0$.

We have found that the p^2 configuration gives rise to the following states:

$$\text{state 1 with } L = 1,\ S = 1$$

$$\text{state 2 with } L = 2,\ S = 0$$

$$\text{state 3 with } L = 0,\ S = 0$$

The total orbital angular momentum in many-electron atoms is designated by the same letters used in one-electron theory. That is, if $L = 0$, we have an S state; if $L = 1$, a P state; and so on. To summarize,

if $L =$	0	1	2	3	4	5	...
the symbol is	S	P	D	F	G	H	...

The total spin angular momentum of the states is designated by a pre-superscript called the multiplicity. The multiplicity is given by $2S + 1$. Thus, we have the states ${}^3P$, ${}^1D$, ${}^1S$, which are read "triplet-P", "singlet-D", and "singlet-S". When the multiplicity is 2, 4, 5, . . . , the states are called doublet, quartet, quintet, etc.

Exercise 6-1:

Construct a microstate chart for the d^2 configuration and determine the Russell-Saunders states which arise. The answer to this problem is given in Appendix 6A.

The ground electronic states of these configurations can be determined from *Hund's first rule* which says that the state with maximum spin multiplicity lies lowest in energy. If there are several states with the same multi-

Table 6-1
Russell-Saunders Terms for Ions with Configurations d^2–d^8

d^2, d^8	${}^1S$, ${}^1D$, ${}^1G$, ${}^3P$, ${}^3F$
d^3, d^7	${}^2D$, ${}^2P$, ${}^2D$, ${}^2F$, ${}^2G$, ${}^2H$, ${}^4P$, ${}^4F$
d^4, d^6	${}^1S$, ${}^1D$, ${}^1G$, ${}^3P$, ${}^3F$, ${}^1S$, ${}^1D$, ${}^1F$, ${}^1G$, ${}^1I$, ${}^3P$, ${}^3D$, ${}^3F$, ${}^3G$, ${}^3H$, ${}^5D$
d^5	${}^2D$, ${}^2P$, ${}^2D$, ${}^2F$, ${}^2G$, ${}^2H$, ${}^4P$, ${}^4F$, ${}^2S$, ${}^2D$, ${}^2F$, ${}^2G$, ${}^2I$ ${}^4D$, ${}^4G$, ${}^6S$

plicity, then *Hund's second rule* says that of these states with the same multiplicity, the one with maximum L lies lowest. Hund's rules can not be used to order the excited states. Thus, for the p^2 configuration the ground state is ^3P, and for the d^2 configuration, the ground state is ^3F.

The Russell-Saunders terms which arise from the configurations d^2 through d^8 are summarized in Table 6.1.

6-3 SPLITTINGS OF RUSSELL-SAUNDERS TERMS IN CRYSTAL FIELDS

In simple crystal field theory it is assumed that the free metallic ion with its atomic energy levels is placed in an electric field set up by the ligand atoms. These ligand atoms perturb the free ion energy levels with the most obvious result being a removal of orbital degeneracy. The magnitude of the energy splittings between the levels that result is determined in large measure by the strength of the crystal field, while the number of levels depends on the symmetry of the crystal field. In this section we will use some group theoretical techniques developed earlier and investigate the splittings of Russell-Saunders terms in crystal fields.

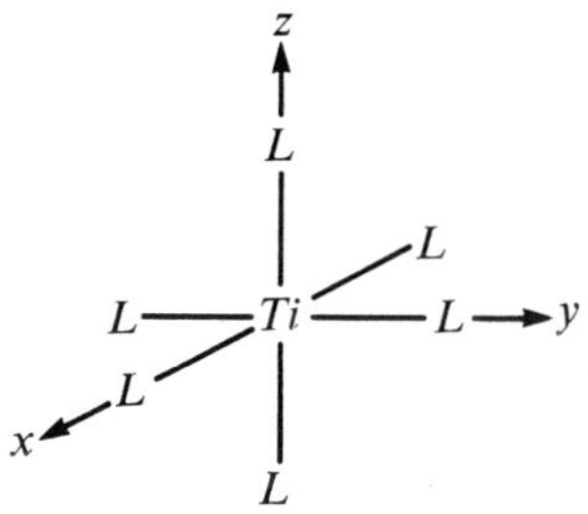

Figure 6.4
Schematic Representation of the Structure of $[TiL_6]^{q\pm}$ *Ion*

It turns out that in most cases the rotational symmetry will suffice to determine the symmetries of the crystal field states. The characters of the representation of a rotational group spanned by an orbital wave function characterized by the quantum number L (or l) are given by

$$\chi(C_\alpha) = \frac{\sin (L + \frac{1}{2})\alpha}{\sin (\alpha/2)} \tag{6-7}$$

where α is the number of degrees of rotation in the operation C_α. The character of the identity operation E is $2L + 1$ (or $2l + 1$). Therefore, for a d^1 system such as the Ti^{3+} ion in an octahedral crystal field shown in Figure 6.4, we must determine the effect of the crystal field symmetry on the ^2D term, since for one d electron the Russell-Saunders term is ^2D.

Now let us construct the representation spanned by the D term in the O-rotational group. In the point group O the operations are C_4, C_3, and C_2. Thus

$$\chi(C_4) = \frac{\sin(2 + \frac{1}{2})\,90^\circ}{\sin 45^\circ} = -1$$

$$\chi(C_3) = \frac{\sin\,(5/2)\,120^\circ}{\sin 60^\circ} = -1$$

$$\chi(C_2) = \frac{\sin\,(5/2)\,180^\circ}{\sin 90^\circ} = 1$$

and

O	E	$8C_3$	$6C_2$	$6C_4$	$3C_2(=C_4^2)$
Γ_D	5	−1	1	−1	1

This reducible representation can be decomposed to $E + T_2$ and these become $E_g + T_{2g}$ in O_h since d orbitals have a center of inversion. That two states arise from ^2D in O_h is all we can learn from symmetry arguments. In order to find out which of the two is the ground state, or what the separation in energy is, we must carry out the calculation described in Section 5 of this chapter.

The splitting of Russell-Saunders terms which arise from any configuration may be determined by constructing the representation in the appropriate rotation group and decomposing the representation by the usual methods. For d^n configurations in octahedral crystal fields all states transform as g, since the d-orbitals are even under the inversion operation. The splittings of several Russell-Saunders terms are given in Table 6.2. In Table 6.2 subscripts of g and u are affixed to the Russell-Saunders term symbol to indicate the parity, that is, whether $\Sigma\, l_i$ is even or odd. For example, an F state arising from one f electron has $\Sigma\, l_i = 3$, and from Table 6.2 we find in the F_u column that, in an O_h field, that the states $A_{2u} + T_{1u} + T_{2u}$ arise. However, an F state which arises from a d^2 configuration, for which $\Sigma\, l_i = 4$, gives $A_{2g} + T_{1g} + T_{2g}$ in an O_h environment.

Table 6-2

Splitting of Russell-Saunders Terms in Crystal Fields of Various Symmetries

Point Group	*Russell-Saunders Term* S_g	S_u	P_g	P_u	D_g	D_u	F_g	F_u	G_g	G_u
O_h	A_{1g}	A_{1u}	T_{1g}	T_{1u}	E_g T_{2g}	E_u T_{2u}	A_{2g} T_{1g} T_{2g}	A_{2u} T_{1u} T_{2u}	A_{1g} E_g T_{1g} T_{2g}	A_{1u} E_u T_{1u} T_{2u}
T_d	A_1	A_2	T_1	T_2	E T_2	E T_1	A_2 T_1 T_2	A_1 T_1 T_2	A_1 E T_1 T_2	A_2 E T_1 T_2
D_{4h}	A_{1g}	A_{1u}	A_{2g} E_g	A_{2u} E_u	A_{1g} B_{1g} B_{2g} E_g	A_{1u} B_{1u} B_{2u} E_u	A_{2g} B_{1g} B_{2g} $2E_g$	A_{2u} B_{1u} B_{2u} $2E_u$	$2A_{1g}$ A_{2g} B_{1g} B_{2g} $2E_g$	$2A_{1u}$ A_{2u} B_{1u} B_{2u} $2E_u$
C_{4v}	A_1	A_2	A_2 E	A_1 E	A_1 B_1 B_2 E	A_2 B_1 B_2 E	A_2 B_1 B_2 $2E$	A_1 B_1 B_2 $2E$	$2A_1$ A_2 B_1 B_2 $2E$	$2A_2$ A_1 B_1 B_2 $2E$
D_{3h}	A'_1	A''_1	A'_2 E''	A''_2 E'	A'_1 E' E''	A''_1 E' E''	A''_1 A'_2 A''_2 E' E''	A'_1 A'_2 A''_2 E' E''	A'_1 A''_1 A''_2 $2E'$ E''	A'_1 A''_1 A'_2 E' $2E''$
C_{3v}	A_1	A_2	A_2 E	A_1 E	A_1 $2E$	A_2 $2E$	A_1 $2A_2$ $2E$	$2A_1$ A_2 $2E$	$2A_1$ A_2 $3E$	A_1 $2A_2$ $3E$
D_{2d}	A_1	B_1	A_2 E	B_2 E	A_1 B_1 B_2 E	A_1 A_2 B_1 E	A_2 B_1 B_2 $2E$	A_1 A_2 B_2 $2E$	$2A_1$ A_2 B_1 B_2 $2E$	A_1 A_2 $2B_1$ B_2 $2E$

Table 6-2 *(continued)*

Splitting of Russell-Saunders Terms in Crystal Fields of Various Symmetries

Point Group	Russell-Saunders Term									
	S_g	S_u	P_g	P_u	D_g	D_u	F_g	F_u	G_g	G_u
D_{2h}	A_g	A_u	B_{1g}	B_{1u}	$2A_g$	$2A_u$	A_g	A_u	$3A_g$	$3A_u$
(D_2)*			B_{2g}	B_{2u}	B_{1g}	B_{1u}	$2B_{1g}$	$2B_{1u}$	$2B_{1g}$	$2B_{1u}$
			B_{3g}	B_{3u}	B_{2g}	B_{2u}	$2B_{2g}$	$2B_{2u}$	$2B_{2g}$	$2B_{2u}$
					B_{3g}	B_{3u}	$2B_{3g}$	$2B_{3u}$	$2B_{3g}$	$2B_{3u}$
C_{2v}	A_1	A_2	A_2	A_1	$2A_1$	A_1	A_1	$2A_1$	$3A_1$	$2A_1$
			B_1	B_1	A_2	$2A_2$	$2A_2$	A_2	$2A_2$	$3A_2$
			B_2	B_2	B_1	B_1	$2B_1$	$2B_1$	$2B_1$	$2B_1$
					B_2	B_2	$2B_2$	$2B_2$	$2B_2$	$2B_2$
C_s	A'	A''	A'	$2A'$	$3A'$	$2A'$	$3A'$	$4A'$	$5A'$	$4A'$
			$2A''$	A''	$2A''$	$3A''$	$4A''$	$3A''$	$4A''$	$5A''$

*For this point group omit *g*, *u* subscripts.

6-4 ONE ELECTRON CORRELATION DIAGRAMS

Since we know the shapes of the *d*-orbitals, and that d_{z^2} and $d_{x^2-y^2}$ transform as E_g while d_{xz}, d_{yz}, and d_{xy} transform as T_{2g}, we can deduce the ordering of the energy levels from simple electrostatic considerations. The electron is intrinsically negatively charged, and the ligands are anions or have the negative ends of their dipoles oriented toward the cation. Therefore, the state which arises when the electron is in one of the two orbitals which points directly toward the ligands should be of higher energy than the state with configuration $(t_{2g})^1$, i.e. with the electron in one of the orbitals d_{xz}, d_{yz}, d_{xy} since these orbitals do not point directly toward the ligands. The energy level diagram is shown in Figure 6.5.

Now let us permit a $[TiL_6]^{q\pm}$ complex to undergo a structural distortion such that the internuclear distance for the two ligands on the *z*-axis in Figure 6.4 is shorter than that for the four ligands in the *xy* plane. The resultant complex belongs to the point group D_{4h}. The most obvious effect of this structural change is a removal of orbital degeneracy. In D_{4h}, d_{z^2} transforms as A_{1g}, and $d_{x^2-y^2}$ as B_{1g}, while the set d_{xz}, d_{yz} forms the basis for E_g and d_{xy} for B_{2g}. A correlation diagram which summarizes this information is given in Figure 6.6. From electrostatic considerations we predict

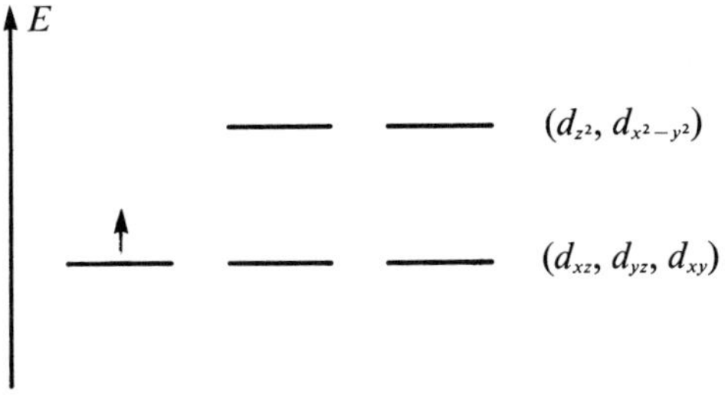

Figure 6.5
Relative Energies of d-Orbitals in O_h *Crystal Fields*

that d_{z^2} will be destabilized with respect to $d_{x^2-y^2}$, and that d_{xz}, d_{yz} will be destabilized with respect to d_{xy}. As we shall see later the splittings of the orbital degeneracies have a considerable effect on such properties as magnetism and optical spectra.

Simple considerations of the sort described above are very useful, but we emphasize that what we have done holds only for one *d*-electron. In Section 6-7 we will develop a procedure for dealing with many-electron ions.

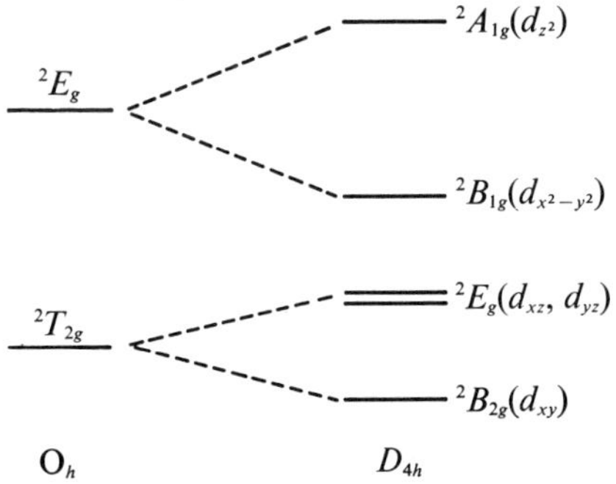

Figure 6.6
Correlation Diagram Showing the Effect of a Tetragonal Distortion on the Octahedral Energy Levels of a ^2D *State*

In Figure 6.7 the one-electron correlation diagrams for a number of chemically interesting geometries are given.

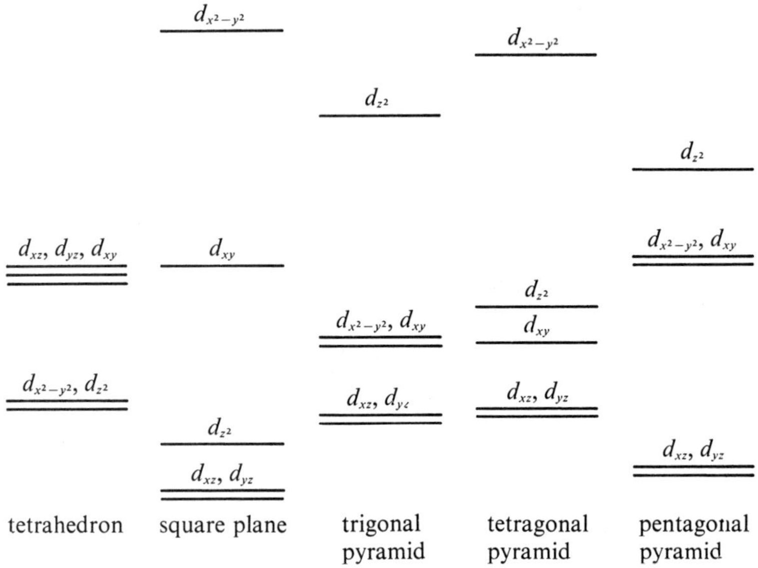

Figure 6.7
One-Electron Correlation Diagrams for Chemically Interesting Structures

6-5 CRYSTAL FIELD CALCULATION FOR ONE ELECTRON

In crystal field theory complexes are treated as though the bonds between the central metal ion and the surrounding ligands are completely ionic. That is, it is assumed that there is no overlap between the metal wave functions and the ligand wave functions. This assumption is admittedly contrary to fact, but the theory is modified later to remove this objection.

For subsequent discussions it is convenient to define the coordinate system shown in Figure 6.8. In this diagram we shall let i label the ligands and j the electrons. It can be shown that the reciprocal of distance between two particles, in our case the electron and one ligand, can be expressed as

$$\frac{1}{r_{ij}} = \sum_{l=0}^{\infty} \sum_{m=-l}^{+l} \frac{4\pi}{2l+1} \frac{r_<^l}{r_>^{l+1}} Y_l^m(\vartheta_i, \varphi_i) Y_l^{m*}(\vartheta_j, \varphi_j) \qquad \textbf{(6-8)}$$

In this equation, the symbols $Y_l^m(\vartheta_i, \varphi_i)$ are the spherical harmonics, some of which are given in Table 6.3. The symbol $r_<$ stands for the lesser of the two distances r_i and r_j while $r_>$ is the greater of the two. This formula

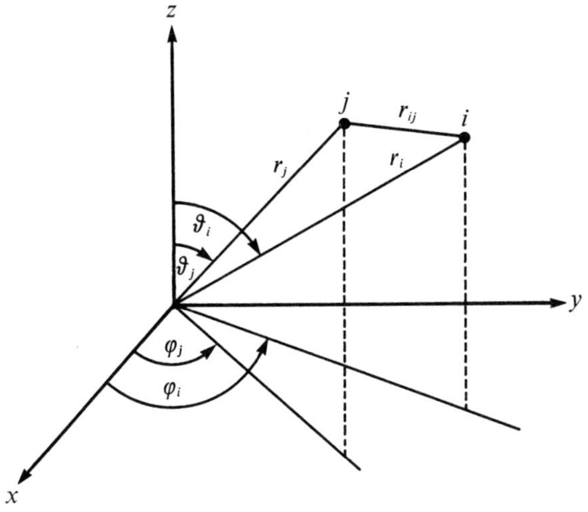

Figure 6.8
Coordinate System for Crystal Field Calculation

is used to construct the crystal field potential energy operator, where $r_<$ pertains to the electron and $r_>$ is the metal-ligand internuclear distance.

The Crystal Field Potential Energy Operator

In crystal field theory the ligands provide a constant electric field that possesses the symmetry of the arrangement of the ligand nuclei. The Hamiltonian is taken to be $H = H_F + V$ where H_F is the Hamiltonian for the free ion and V is the crystal field potential energy operator. It is assumed that the eigenvalues and eigenfunctions of the free ion are known, and V is regarded as a perturbation. This approach is called the weak crystal field approximation. Strong-field and intermediate-field calculations will not be covered here.

Here we will investigate the nature of V for the octahedral complex shown in Figure 6.4. Using Equation (6-8), $V(O_h)$ is written as

$$V(O_h) = \sum_i \sum_j \sum_{l=0}^{\infty} \sum_{m=-l}^{+l} Z_q e^2 \frac{4\pi}{2l+1} \frac{r_<^l}{r_>^{l+1}} Y_{l_j}^m Y_{l_i}^{m*} \tag{6-9}$$

where the summations are over the six ligands j and the electrons i. Only a few terms in the summation are actually used to make up $V(O_h)$. All terms with l equal to an odd number are neglected as are all terms with

Table 6-3

A. Spherical Harmonics in Rectangular Coordinates

$$Y_0^0 = \sqrt{\frac{1}{4\pi}}$$

$$Y_1^{-1} = \sqrt{\frac{3}{8\pi}}\frac{x - iy}{r}$$

$$Y_1^0 = \sqrt{\frac{3}{4\pi}}\frac{z}{r}$$

$$Y_1^1 = -\sqrt{\frac{3}{8\pi}}\frac{x + iy}{r}$$

$$Y_2^{-2} = \sqrt{\frac{5}{4\pi}}\sqrt{\frac{3}{8}}\frac{(x - iy)^2}{r^2}$$

$$Y_2^{-1} = \sqrt{\frac{5}{4\pi}}\sqrt{\frac{3}{2}}\frac{z(x - iy)}{r^2}$$

$$Y_2^0 = \sqrt{\frac{5}{4\pi}}\sqrt{\frac{1}{4}}\frac{3z^2 - r^2}{r^2}$$

$$Y_2^1 = -\sqrt{\frac{5}{4\pi}}\sqrt{\frac{3}{2}}\frac{z(x + iy)}{r^2}$$

$$Y_2^2 = \sqrt{\frac{5}{4\pi}}\sqrt{\frac{3}{8}}\frac{(x + iy)^2}{r^2}$$

$$Y_3^{-3} = \sqrt{\frac{7}{4\pi}}\sqrt{\frac{5}{16}}\frac{(x - iy)^3}{r^3}$$

$$Y_3^{-2} = \sqrt{\frac{7}{4\pi}}\sqrt{\frac{15}{8}}\frac{z(x - iy)^2}{r^3}$$

$$Y_3^{-1} = \sqrt{\frac{7}{4\pi}}\sqrt{\frac{3}{16}}\frac{(x - iy)(5z^2 - r^2)}{r^3}$$

$$Y_3^0 = \sqrt{\frac{7}{4\pi}}\sqrt{\frac{1}{4}}\frac{z(5z^2 - 3r^2)}{r^3}$$

$$Y_3^1 = -\sqrt{\frac{7}{4\pi}}\sqrt{\frac{3}{16}}\frac{(x + iy)(5z^2 - r^2)}{r^3}$$

$$Y_3^2 = \sqrt{\frac{7}{4\pi}}\sqrt{\frac{15}{8}}\frac{z(x + iy)^2}{r^3}$$

$$Y_3^3 = -\sqrt{\frac{7}{4\pi}}\sqrt{\frac{5}{16}}\frac{(x + iy)^3}{r^3}$$

Table 6-3 *(continued)*

$$Y_4^{-4} = \sqrt{\frac{9}{4\pi}}\sqrt{\frac{35}{128}}\frac{(x - iy)^4}{r^4}$$

$$Y_4^{-3} = \sqrt{\frac{9}{4\pi}}\sqrt{\frac{35}{16}}\frac{z(x - iy)^3}{r^4}$$

$$Y_4^{-2} = \sqrt{\frac{9}{4\pi}}\sqrt{\frac{5}{32}}\frac{(x - iy)^2}{r^4}(7z^2 - r^2)$$

$$Y_4^{-1} = \sqrt{\frac{9}{4\pi}}\sqrt{\frac{5}{16}}\frac{(x - iy)}{r^4}(7z^3 - 3zr^2)$$

$$Y_4^{0} = \sqrt{\frac{9}{4\pi}}\sqrt{\frac{1}{64}}\frac{35z^4 - 30z^2r^2 + 3r^4}{r^4}$$

$$Y_4^{1} = -\sqrt{\frac{9}{4\pi}}\sqrt{\frac{5}{16}}\frac{(x + iy)}{r^4}(7z^3 - 3zr^2)$$

$$Y_4^{2} = \sqrt{\frac{9}{4\pi}}\sqrt{\frac{5}{32}}\frac{(x + iy)^2}{r^4}(7z^2 - r^2)$$

$$Y_4^{3} = -\sqrt{\frac{9}{4\pi}}\sqrt{\frac{35}{16}}\frac{z(x + iy)^3}{r^4}$$

$$Y_4^{4} = \sqrt{\frac{9}{4\pi}}\sqrt{\frac{35}{128}}\frac{(x + iy)^4}{r^4}$$

$$Y_5^{-5} = \sqrt{\frac{11}{4\pi}}\sqrt{\frac{63}{256}}\frac{(x - iy)^5}{r^5}$$

$$Y_5^{-4} = \sqrt{\frac{11}{4\pi}}\sqrt{\frac{315}{128}}\frac{z(x - iy)^4}{r^5}$$

$$Y_5^{-3} = \sqrt{\frac{11}{4\pi}}\sqrt{\frac{35}{256}}\frac{(x - iy)^3}{r^5}(9z^2 - r^2)$$

$$Y_5^{-2} = \sqrt{\frac{11}{4\pi}}\sqrt{\frac{105}{32}}\frac{(x - iy)^2}{r^5}(3z^3 - zr^2)$$

$$Y_5^{-1} = \sqrt{\frac{11}{4\pi}}\sqrt{\frac{15}{128}}\frac{(x - iy)}{r^5}(21z^4 - 14z^2r^2 + r^4)$$

$$Y_5^{0} = \sqrt{\frac{11}{4\pi}}\frac{1}{8}\frac{63z^5 - 70z^3r^2 + 15zr^4}{r^5}$$

$$Y_5^{1} = -\sqrt{\frac{11}{4\pi}}\sqrt{\frac{15}{128}}\frac{x + iy}{r^5}(21z^4 - 14z^2r^2 + r^4)$$

$$Y_5^{2} = \sqrt{\frac{11}{4\pi}}\sqrt{\frac{105}{32}}\frac{(x + iy)^2}{r^5}(3z^3 - zr^2)$$

Table 6-3 *(continued)*

$$Y_5^3 = -\sqrt{\frac{11}{4\pi}}\sqrt{\frac{35}{256}}\frac{(x+iy)^3}{r^5}(9z^2 - r^2)$$

$$Y_5^4 = \sqrt{\frac{11}{4\pi}}\sqrt{\frac{315}{128}}\frac{z(x+iy)^4}{r^5}$$

$$Y_5^5 = -\sqrt{\frac{11}{4\pi}}\sqrt{\frac{63}{256}}\frac{(x+iy)^5}{r^5}$$

Reprinted with permission from C. J. Ballhausen, *Introduction to Ligand Field Theory*, New York: McGraw-Hill Book Co., Inc., 1962.

B. Spherical Harmonics in Spherical Polar Coordinates

$$Y_l^m = \Phi_{m_l}(\varphi)\Theta_{lm_l}(\vartheta)$$

The Functions $\Phi_m(\varphi)$

$$\Phi_0(\varphi) = \frac{1}{\sqrt{2\pi}} \quad \text{or} \quad \Phi_0(\varphi) = \frac{1}{\sqrt{2\pi}}$$

$$\Phi_1(\varphi) = \frac{1}{\sqrt{2\pi}}e^{i\varphi} \quad \text{or} \quad \Phi_{1\,\cos}(\varphi) = \frac{1}{\sqrt{\pi}}\cos\varphi$$

$$\Phi_{-1}(\varphi) = \frac{1}{\sqrt{2\pi}}e^{-i\varphi} \quad \text{or} \quad \Phi_{1\,\sin}(\varphi) = \frac{1}{\sqrt{\pi}}\sin\varphi$$

$$\Phi_2(\varphi) = \frac{1}{\sqrt{2\pi}}e^{i2\varphi} \quad \text{or} \quad \Phi_{2\,\cos}(\varphi) = \frac{1}{\sqrt{\pi}}\cos 2\varphi$$

$$\Phi_{-2}(\varphi) = \frac{1}{\sqrt{2\pi}}e^{-i2\varphi} \quad \text{or} \quad \Phi_{2\,\sin}(\varphi) = \frac{1}{\sqrt{\pi}}\sin 2\varphi$$

Etc.

The Functions θ_{lm}

$l = 0$

$$\theta_{00}(\vartheta) = \frac{\sqrt{2}}{2}$$

$l = 1$

$$\theta_{10}(\vartheta) = \frac{\sqrt{6}}{2}\cos\vartheta$$

$$\theta_{1\pm1}(\vartheta) = \frac{\sqrt{3}}{2}\sin\vartheta$$

Table 6-3 *(continued)*

$l = 2$

$$\theta_{20}(\vartheta) = \frac{\sqrt{10}}{4}(3\cos^2\vartheta - 1)$$

$$\theta_{2\pm1}(\vartheta) = \frac{\sqrt{15}}{2}\sin\vartheta\cos\vartheta$$

$$\theta_{2\pm2}(\vartheta) = \frac{\sqrt{15}}{4}\sin^2\vartheta$$

$l = 3$

$$\theta_{30}(\vartheta) = \frac{3\sqrt{14}}{4}\left(\frac{5}{3}\cos^3\vartheta - \cos\vartheta\right)$$

$$\theta_{3\pm1}(\vartheta) = \frac{\sqrt{42}}{8}\sin\vartheta(5\cos^2\vartheta - 1)$$

$$\theta_{3\pm2}(\vartheta) = \frac{\sqrt{105}}{4}\sin^2\vartheta\cos\vartheta$$

$$\theta_{3\pm3}(\vartheta) = \frac{\sqrt{70}}{8}\sin^2\vartheta$$

$l = 4$

$$\theta_{40}(\vartheta) = \frac{9\sqrt{2}}{16}\left(\frac{35}{3}\cos^4\vartheta - 10\cos^2\vartheta + 1\right)$$

$$\theta_{4\pm1}(\vartheta) = \frac{9\sqrt{10}}{8}\sin\vartheta\left(\frac{7}{3}\cos^3\vartheta - \cos\vartheta\right)$$

$$\theta_{4\pm2}(\vartheta) = \frac{3\sqrt{5}}{8}\sin^2\vartheta\,(7\cos^2\vartheta - 1)$$

$$\theta_{4\pm3}(\vartheta) = \frac{3\sqrt{70}}{8}\sin^3\vartheta\cos\vartheta$$

$$\theta_{4\pm4}(\vartheta) = \frac{3\sqrt{35}}{16}\sin^4\vartheta$$

$l = 5$

$$\theta_{50}(\vartheta) = \frac{15\sqrt{22}}{16}\left(\frac{21}{5}\cos^5\vartheta - \frac{14}{3}\cos^3\vartheta + \cos\vartheta\right)$$

$$\theta_{5\pm1}(\vartheta) = \frac{\sqrt{165}}{16}\sin\vartheta(21\cos^4\vartheta - 14\cos^2\vartheta + 1)$$

$$\theta_{5\pm2}(\vartheta) = \frac{\sqrt{1155}}{8}\sin^2\vartheta(3\cos^3\vartheta - \cos\vartheta)$$

Table 6-3 *(continued)*

$$\Theta_{5\pm3}(\vartheta) = \frac{\sqrt{770}}{32}\sin^3\vartheta(9\cos^2\vartheta - 1)$$

$$\Theta_{5\pm4}(\vartheta) = \frac{3\sqrt{385}}{16}\sin^4\vartheta\cos\vartheta$$

$$\Theta_{5\pm5}(\vartheta) = \frac{3\sqrt{154}}{32}\sin^5\vartheta$$

l greater than four. (One can show that integrals of the sort

$$\int_0^{\pi}\int_0^{2\pi} Y_l^{m*}Y_{l'}^{m'}Y_l^{m''}\sin\theta\, d\theta\, d\phi = 0$$

when $l > 2l'$. The mathematical proof is tedious and will not be given here.)

To determine the terms kept in the potential, we make note of the fact that the potential energy operator must transform as the totally symmetric representation of the group. Thus, we determine the representation spanned by the sets Y_l^m and retain only those that transform as A_{1g} in O_h.

The first term in the expansion is spherically symmetric, transforms as A_{1g}, but does not remove the orbital degeneracy of the d orbitals. This term is very important in determining lattice and hydration energies, but it is not necessary to carry this term along in most crystal field considerations.

The second set of terms in the expansion, i.e. those with $l = 1$ and $m = 1, 0, -1$ can be shown by application of the formula $[\sin(l + \frac{1}{2})\alpha]/(\sin\alpha/2)$ to form the following representation of the rotation group O:

O	E	$8C_3$	$3C_2$	$6C_4$	$6C_2'$	
$\Gamma_{Y_1^m}$	3	0	−1	1	−1	$= T_1$

Since the set Y_1^m transforms as T_1 and not as A_1, these terms do not enter into the expression for $V(O_h)$.

By similar methods it can be seen that the set Y_2^m transforms as $E + T_2$ while the set Y_3^m forms the basis for $A_2 + T_1 + T_2$. Finally, the set Y_4^m transforms as $A_1 + E + T_1 + T_2$. Thus, a linear combination of some Y_4^m make up the crystal field potential.

In order to determine these terms we choose the C_4 axis as the axis of quantization and find those terms which are invariant to the operation C_4. Using the spherical harmonics written in terms of x, y, and z given in

Table 6.3, we find

$$C_4 \begin{bmatrix} Y_4^4 \\ Y_4^3 \\ Y_4^2 \\ Y_4^1 \\ Y_4^0 \\ Y_4^{-1} \\ Y_4^{-2} \\ Y_4^{-3} \\ Y_4^{-4} \end{bmatrix} = \begin{bmatrix} Y_4^4 \\ iY_4^3 \\ -Y_4^2 \\ -iY_4^1 \\ Y_4^0 \\ iY_4^{-1} \\ -Y_4^{-2} \\ -iY_4^{-3} \\ Y_4^{-4} \end{bmatrix}$$

To illustrate how the above results are obtained, examine the Y_4^4 case. The function for Y_4^4 is obtained from Table 6.3; it is $Y_4^4 = (x + iy)^4$. Now operate on this function with C_4. The result is $C_4 Y_4^4 = [y + i(-x)]^4$, since the C_4 operation takes x into y and y into $-x$. If we multiply y by $(-i)(i)$ (which is 1) we get

$$C_4 Y_4^4 = [i(-i)y - ix]^4 = (-i)^4 (x + iy)^4 = (x + iy)^4 = Y_4^4$$

Since $V(O_h)$ must transform as A_{1g} we have $V_4 = aY_4^4 + bY_4^0 + cY_4^{-4}$

or
$$V_4 = Y_4^0 + a/b\ Y_4^4 + c/b\ Y_4^{-4} \tag{6-10}$$

(Y_4^0, $Y_4^{\pm 4}$ were the only terms invariant under the C_4 operation.)

To evaluate the constants a, b, c, we see what happens to Y_4^4, Y_4^0, and Y_4^{-4} under some other operations of the group. Choose C_2' as the first symmetry operator. It can be seen that

$$C_2' Y_4^4 = Y_4^{-4} \qquad C_2' Y_4^0 = Y_4^0 \qquad C_2' Y_4^{-4} = Y_4^4$$

Therefore a/b must be equal to c/b, and we have

$$V(O_h) = Y_4^0 + \frac{a}{b}(Y_4^4 + Y_4^{-4}) = \sqrt{\frac{9}{4\pi}}\left[\left(\frac{1}{8}\frac{35z^4 - 30z^2r^2 + 3r^4}{r^4}\right) + \frac{a}{b}\left(\sqrt{\frac{35}{128}}\frac{(x+iy)^4}{r^4} + \sqrt{\frac{35}{128}}\frac{(x-iy)^4}{r^4}\right)\right] \tag{6-11}$$

Application of the C_3 operation to this will permit the evaluation of a/b.

We see that

$$C_3V_4 = \sqrt{\frac{9}{4\pi}}\left[\left(\frac{1}{8}\frac{35x^4 - 30x^2r^2 + 3r^4}{r^4}\right) + \frac{a}{b}\sqrt{\frac{35}{128}}\left(\frac{(y+iz)^4}{r^4} + \frac{(y-iz)^4}{r^4}\right)\right]$$

Since $C_3V_4 = V_4$, we write

$$\frac{1}{8}(35z^4 - 30r^2z^2 + 3r^4) + \frac{a}{b}\sqrt{\frac{35}{128}}[(x+iy)^4 + (x-iy)^4]$$
$$= \frac{1}{8}(35x^4 - 30x^2r^2 + 3r^4) + \frac{a}{b}\sqrt{\frac{35}{128}}[(y+iz)^4 + (y-iz)^4]$$

For the two sides to be equal, the coefficients of like powers of the variables must be equal, so collect terms on z^4. This gives

$$\frac{1}{8}(35z^4 - 30z^4) = 2\frac{a}{b}\sqrt{\frac{35}{128}}z^4 \quad \text{and} \quad \frac{a}{b} = \sqrt{\frac{5}{14}}$$

Therefore

$$V(O_h) = Y_4^0 + \sqrt{5/14}\,(Y_4^4 + Y_4^{-4}) \tag{6-12}$$

Now we complete the expression by carrying out the summation over the ligands which are placed on the x, y, z axes at $\pm a$. The expression for Y_4^0 is

$$Y_4^0 = \sqrt{\frac{9}{128}}\sqrt{\frac{1}{2\pi}}(35\cos^4\vartheta - 30\cos^2\vartheta + 3).$$

The six ligands have the coordinates and ϑ values shown below:

x	y	z	ϑ	$35\cos^4\vartheta - 30\cos^2\vartheta + 3$
0	0	a	0	8
0	0	$-a$	π	8
0	a	0	$\pi/2$	3
0	$-a$	0	$\pi/2$	3
a	0	0	$\pi/2$	3
$-a$	0	0	$\pi/2$	3

The first term is

$$\frac{4\pi}{9}\sqrt{\frac{9}{128}}\sqrt{\frac{1}{2\pi}}(2\times 8 + 4\times 3)\frac{Ze^2r^4}{a^5}\,Y_4^0\,\vartheta_i, \varphi_i \tag{6-13}$$

For the $Y_4^{\pm 4}$ terms, we first note that

$$Y_4^{\pm 4} = \sqrt{\frac{315}{256}}\sqrt{\frac{1}{2\pi}} \sin^4 \vartheta e^{\pm 4i\varphi}$$

The coordinates, ϑ and φ values are

x	y	z	ϑ	φ	$\sin^4 \vartheta e^{\pm 4i\varphi}$
0	0	a	0	—	0
0	0	$-a$	π	—	0
0	a	0	$\pi/2$	$\pi/2$	1
0	$-a$	0	$\pi/2$	$-\pi/2$	1
a	0	0	$\pi/2$	0	1
$-a$	0	0	$\pi/2$	π	1

and the second term is

$$\frac{4\pi}{9}\sqrt{\frac{1}{2\pi}}\sqrt{\frac{315}{256}}(4Y_4^4 + 4Y_4^{-4})\frac{Ze^2r^4}{a^5} \qquad \text{(6-14)}$$

Combining the two terms in Equations (6-13) and (6-14) the total potential becomes

$$V(\mathrm{O}_h) = \frac{7}{3}\pi^{1/2}\frac{Ze^2r^4}{a^5}\left[Y_4^0 + \sqrt{\frac{5}{14}}(Y_4^4 + Y_4^{-4})\right] \qquad \text{(6-15)}$$

which in Cartesian form is

$$V(\mathrm{O}_h) = \frac{35Ze^2}{4a^5}\left(x^4 + y^4 + z^4 - \frac{3}{5}r^4\right) \qquad \text{(6-16)}$$

Exercise 6-2:

Show that Equation (6-15) is equivalent to Equation (6-16).

The Calculation

According to degenerate perturbation theory we must evaluate the elements in the equation $|H_{ij} - ES_{ij}| = 0$ in order to obtain the energy levels which result. Since we are dealing with an orthogonal set of orbitals, then $S_{ii} = 1$ and $S_{ij} = 0$. We must evaluate integrals of the form $\int \psi_i^* V \psi_j d\tau$, which we will designate as $\langle\psi_i|V|\psi_j\rangle$. Since we are using the weak-field approximation, we choose as the basis set of wave functions for the calculation the hydrogenic d-functions which may be written as $\psi = NR(r)\,\Theta(\vartheta)\Phi(\varphi)$. Integrals of the type $\langle R(r)\,\Theta(\vartheta)\Phi(\varphi)\,|\,V(r, \vartheta, \varphi)|\,R(r)\,\Theta(\vartheta)\Phi(\varphi)\rangle$ must be evaluated. We will first investigate the ϕ dependent part of the problem. This involves only $x^4 + y^4$. Since $x = r \sin\vartheta \cos\varphi$ and $y = r \sin\vartheta \sin\varphi$, then $x^4 + y^4 = r^4 \sin^4\vartheta\,(\sin^4\varphi + \cos^4\varphi)$. The integrals will be of the type $\langle e^{-im'\varphi}|\sin^4\varphi + \cos^4\varphi\,|\,e^{im\varphi}\rangle$. Upon substitution for the operator we get

$$\left\langle e^{-im'\varphi}\left|\left(\frac{e^{i\varphi} - e^{-i\varphi}}{2i}\right)^4 + \left(\frac{e^{i\varphi} + e^{-i\varphi}}{2}\right)^4\right|e^{im\varphi}\right\rangle$$

This simplifies to

$$\langle e^{-im'\varphi}\,|\,\tfrac{1}{8}(e^{4i\varphi} + e^{-4i\varphi} + 6)\,|\,e^{im\varphi}\rangle$$

where m and m' take on values of 0, ± 1, ± 2.

Let us look at a typical integral: $\int_0^{2\pi} e^{-im\varphi}\, e^{im'\varphi}\, e^{im''\varphi}\, d\varphi = \int_0^{2\pi} e^{im'''\varphi}\, d\tau$. This integral can be nonzero only if $m''' = 0$. Consequently, in the determinantal equation only those matrix elements designated by $\langle\ \rangle$ will be nonzero:

$$\begin{array}{c} \\ 2 \\ 1 \\ 0 \\ -1 \\ -2 \end{array}\begin{array}{c} \begin{array}{ccccc} 2 & 1 & 0 & -1 & -2 \end{array} \\ \left|\begin{array}{ccccc} \langle\ \rangle & 0 & 0 & 0 & \langle\ \rangle \\ 0 & \langle\ \rangle & 0 & 0 & 0 \\ 0 & 0 & \langle\ \rangle & 0 & 0 \\ 0 & 0 & 0 & \langle\ \rangle & 0 \\ \langle\ \rangle & 0 & 0 & 0 & \langle\ \rangle \end{array}\right| = 0 \end{array}$$

We must now calculate the nonzero matrix elements using the φ-dependent part of the potential energy operator. A typical integral is

$$\langle e^{-2i\varphi}\,|\,V(\varphi)\,|\,e^{2i\varphi}\rangle$$

where

$$\begin{aligned} V(\varphi) &= x^4 + y^4 \\ &= r^4 \sin^4\vartheta\,(\sin^4\varphi + \cos^4\varphi) \\ &= \tfrac{1}{8}\, r^4 \sin^4\vartheta\,(e^{4i\varphi} + e^{-4i\varphi} + 6) \end{aligned}$$

Substitution for the operator yields

$$\tfrac{1}{8} r^4 \sin^4 \vartheta \langle e^{-2i\varphi} | e^{4i\varphi} + e^{-4i\varphi} + 6 | e^{2i\varphi} \rangle$$
$$= \tfrac{3}{4} r^4 \sin^4 \vartheta \langle e^{-2i\varphi} | e^{2i\varphi} \rangle$$
$$= \tfrac{3}{4} r^4 \sin^4 \vartheta$$

The values for the non-zero matrix elements are:

$m:m'$		
$2:2$ $1:1$ $0:0$ $-1:-1$ $-2:-2$	$\langle \Phi(m) | V(\phi) | \Phi(m) \rangle = \tfrac{3}{4} r^4 \sin^4 \theta$	**(6-17)**
$2:-2$ $-2:2$	$\langle \Phi(m) | V(\phi) | \Phi(m) \rangle = \tfrac{1}{8} r^4 \sin^4 \theta$	

Now we shall look at the θ-dependent part of the nonzero matrix elements. First recall that we are looking at integrals of the form

$$\int_0^\infty \int_0^\pi \int_0^{2\pi} \Psi^*_{n,l,m}(r, \theta, \phi)[V(r, \theta, \phi)]\Psi_{n,l,m}(r, \theta, \phi) r^2 \sin \theta dr d\theta d\phi$$

where we have given values above for the integrals $\langle \Phi_m | V | \Phi'_m \rangle$. Of course we did not consider the z^4 term since it is not ϕ-dependent. Consequently we must add $r^4 \sin^4 \theta$ to each of the terms in Equation (6-17). As an example consider the integral

$$\langle 0 | V(\theta, \phi) | 0 \rangle = \int_0^\pi \theta_2^{0*} \langle \Phi_0 | V | \Phi_0 \rangle \theta_2^0 \sin \theta \, d\theta$$

This becomes

$$\tfrac{5}{8} \int_0^\pi (3 \cos^2 \theta - 1)(\tfrac{3}{4} r^4 \sin^4 \theta + r^4 \cos^4 \theta)(3 \cos^2 \theta - 1) \sin \theta \, d\theta$$
$$= \tfrac{5}{8} r^4 \int_0^\pi (\tfrac{27}{4} \sin^5 \theta \cos^4 \theta - \tfrac{9}{2} \cos^2 \theta \sin^5 \theta + \tfrac{3}{4} \sin^5 \theta$$
$$+ 9 \sin \vartheta \cos^8 \theta - 6 \sin \theta \cos^6 \theta + \sin \theta \cos^4 \theta) \, d\theta$$

From a mathematical handbook we find that in general

$$\int_0^{\pi/2} \sin^{n-1} \theta \cos^{m-1} \theta \, d\theta = \frac{1}{2} B\left(\frac{n}{2}, \frac{m}{2}\right)$$

where

$$B(j, k) = \frac{\Gamma(j)\Gamma(k)}{\Gamma(j+k)}$$

and

$$\Gamma(j) = (j-1)!$$

$$\Gamma_{(j+1/2)} = \frac{1 \cdot 3 \cdot 5 \cdot 7 \cdots (2j-1)}{2^j} \sqrt{\pi}$$

Noting that only even powers of cos ϑ are involved, we can change the limits on the integral giving

$$\langle 0 | V(\theta, \phi) | 0 \rangle = \tfrac{5}{4} r^4 \int_0^{\pi/2} (\ldots)\, d\theta$$

For the first term

$$\int_0^{\pi/2} \sin^5\theta \cos^4\theta \, d\theta$$

$n = 6$ and $m = 5$. We want $\frac{1}{2} B\left(\frac{n}{2}, \frac{m}{2}\right) = \frac{1}{2} B\left(3, \frac{5}{2}\right)$.

It can be seen that

$$\frac{1}{2} B(3, \tfrac{5}{2}) = \frac{1}{2} \frac{\Gamma(3)\Gamma(2+\frac{1}{2})}{\Gamma(5+\frac{1}{2})} = \frac{8}{5 \cdot 7 \cdot 9}$$

The result is

$$\begin{aligned}\langle 0 | V(\theta, \phi) | 0 \rangle &= \frac{5}{4} r^4 \left[\frac{27}{4} \cdot \frac{8}{5 \cdot 7 \cdot 9} - \frac{9}{2} \cdot \frac{8}{3 \cdot 5 \cdot 7} + \frac{3}{4} \cdot \frac{8}{3 \cdot 5} \right. \\ &\quad \left. + \frac{9}{9} - \frac{6}{7} + \frac{1}{5}\right] \\ &= \frac{5}{7} r^4\end{aligned}$$

The other integrals have the values:

$$\begin{aligned}\langle 1 | V(\theta, \phi) | 1 \rangle &= \langle -1 | V(\theta, \phi) | -1 \rangle = \tfrac{11}{21} r^4 \\ \langle 2 | V(\theta, \phi) | 2 \rangle &= \langle -2 | V(\theta, \phi) | -2 \rangle = \tfrac{13}{21} r^4 \qquad \text{(6-18)} \\ \langle 2 | V(\theta, \phi) | -2 \rangle &= \langle -2 | V(\theta, \phi) | \ 2 \rangle = \tfrac{73}{105} r^4\end{aligned}$$

Completing the evaluation of the nonzero matrix elements by including the radial dependence, we have

$$\langle R(r)|\, D(x^4 + y^4 + z^4 - \tfrac{3}{5} r^4)\,| R(r)\rangle$$

where

$$D = \frac{35Ze^2}{4a^5}$$

As an example we do the integral

$$\begin{aligned}\langle 0|V|0\rangle &= \tfrac{5}{7} D \int_0^\infty R(r)^2 r^4 r^2\, dr - \tfrac{3}{5} D \int_0^\infty R(r)^2 r^4 r^2\, dr \\ &= \tfrac{4}{35} D \int_0^\infty R(r)^2 r^4 r^2\, dr\end{aligned}$$

This result is defined as $6Dq$ where $q = \dfrac{2}{105}\displaystyle\int_0^\infty R(r)^2 r^4 r^2\, dr = \dfrac{2}{105}\langle r^4\rangle$. The results are:

$$\begin{aligned}\langle 1|V|1\rangle &= \langle -1|V|-1\rangle = -4Dq \\ \langle 2|V|2\rangle &= \langle -2|V|-2\rangle = Dq \\ \langle -2|V|2\rangle &= \langle \ 2|V|-2\rangle = 5Dq\end{aligned} \tag{6-19}$$

The perturbation secular determinantal equation for the energy is

$$\begin{vmatrix} Dq - E & 0 & 0 & 0 & 5Dq \\ 0 & -4Dq - E & 0 & 0 & 0 \\ 0 & 0 & 6Dq - E & 0 & 0 \\ 0 & 0 & 0 & -4Dq - E & 0 \\ 5Dq & 0 & 0 & 0 & Dq - E \end{vmatrix} = 0$$

The 1×1 equations give the roots

$$\begin{aligned}E &= -4Dq \\ E &= -4Dq \\ E &= \ \ 6Dq\end{aligned}$$

while the two by two determinantal equation

$$\begin{vmatrix} Dq - E & 5Dq \\ 5Dq & Dq - E \end{vmatrix} = Dq^2 - 2DqE + E^2 - 25Dq = 0$$

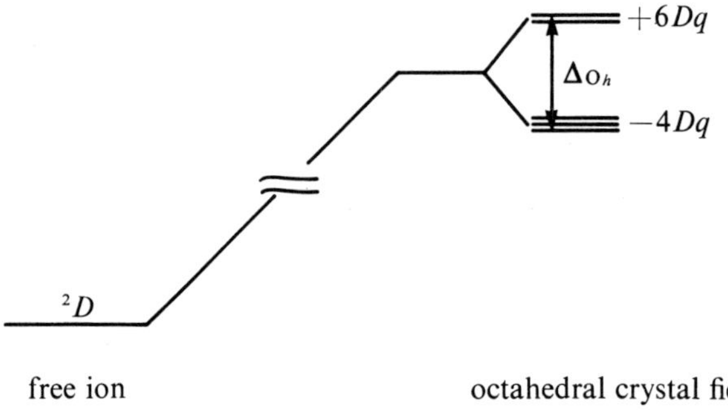

Figure 6.9
Splitting of $^2T_{2g}$ Term in O_h Crystal Field

gives the two roots

$$E = 6Dq$$
$$E = -4Dq$$

Thus, in an octahedral crystal field the ^2D term splits as shown in Figure 6.9.

We can determine the wave functions from the secular equations by substituting in the energies obtained above. For the diagonal terms in the determinant, we find d_0 or d_{z^2} is one of the eigenfunctions corresponding to the eigenvalue $+6Dq$, and that d_{+1} and d_{-1} correspond to two of the roots at $-4Dq$. Thus d_{+2} and d_{-2} must combine to give eigenfunctions for the remaining $6Dq$ and $-4Dq$ eigenvalues. We have the two linear equations:

$$(Dq - E)C_1 + 5Dq\, C_2 = 0$$
$$5Dq\, C_1 + (Dq - E)C_2 = 0$$

When $E = 6Dq$:

$$-5Dq\, C_1 + 5Dq\, C_2 = 0$$
$$5Dq\, C_1 - 5Dq\, C_2 = 0$$

or $C_1 = C_2$, and the function is $(d_2 + d_{-2})/\sqrt{2}$.
When $E = -4Dq$, we have

$$5Dq\, C_1 + 5Dq\, C_2 = 0$$
$$5Dq\, C_1 + 5Dq\, C_2 = 0$$

and $C_1 = -C_2$ giving the function $(d_2 - d_{-2})/\sqrt{2}$.

Radial Parameters and Lower Symmetry Geometries

The energies of the two crystal field states have been expressed in terms of the parameter Dq, where q is proportional to $\langle r^4 \rangle$. Even when the best available atomic radial wave functions are used, only poor agreement between experimental and calculated Dq's is obtained. For this reason Dq is treated as a parameter to be determined from experiment.

For lower symmetry complexes additional parameters arise. Consider a tetragonal distortion along the z-axis of an octahedral complex such that the ligands are now at the positions shown in Figure 6.10. As a result of this distortion the C_3 axis is lost but the C_2 and C_4 axes are retained. In the rotation group D_4 we find

D_4	E	$2C_4$	C_2	$2C_2'$	$2C_2''$	
Y_2^m	5	-1	1	1	1	$= A_1 + \ldots$
Y_4^m	9	1	1	1	1	$= 2A_1 + \ldots$

Therefore in addition to Y_4^m there will be a contribution from Y_2^m in the crystal field potential energy operator.

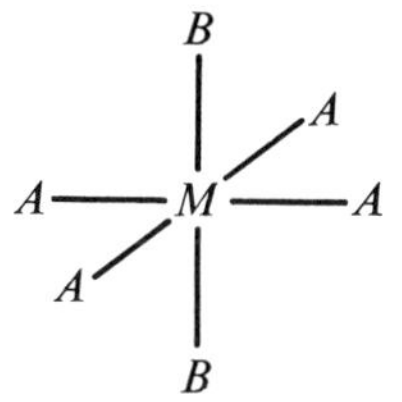

Figure 6.10
Schematic Representation of MA_4B_2 Molecule

In Cartesian coordinates the second order spherical harmonics are

$$
\begin{aligned}
Y_2^2 &= \sqrt{5/4\pi}\,\sqrt{3/8}\,(x + iy)^2/r^2 \\
Y_2^1 &= -\sqrt{5/4\pi}\,\sqrt{3/2}\,(z)(x + iy)/r^2 \\
Y_2^0 &= \sqrt{5/4\pi}\,\sqrt{1/4}\,(3z^2 - r^2)/r^2 \\
Y_2^{-1} &= \sqrt{5/4\pi}\,\sqrt{3/2}\,(z)(x - iy)/r^2 \\
Y_2^{-2} &= \sqrt{5/4\pi}\,\sqrt{3/8}\,(x - iy)^2/r^2
\end{aligned}
\qquad \textbf{(6-20)}
$$

Operating on this set of spherical harmonics with C_4 we get

$$C_4 \begin{bmatrix} Y_2^2 \\ Y_2^1 \\ Y_2^0 \\ Y_2^{-1} \\ Y_2^{-2} \end{bmatrix} = \begin{bmatrix} -Y_2^2 \\ -iY_2^1 \\ Y_2^0 \\ iY_2^{-1} \\ -Y_2^{-2} \end{bmatrix}$$

Since only Y_2^0 is invariant with respect to the quantization axis, only Y_2^0 will be added to the octahedral field potential, and

$$V_{\text{Tetragonal}} = aY_2^0 + bY_4^0 + c(Y_4^4 + Y_4^{-4})$$

The constants a, b, c can be determined in a manner identical to that used before to determine the constants for the octahedral crystal field potential. The result is

$$V_{\text{Tetragonal}} = (2)\sqrt{2\pi}\, Ze^2 \left\{-\left(\frac{2}{5}\right)^{1/2} r^2(a^{-3} - b^{-3})Y_2^0 + \left(\frac{1}{72}\right)^{1/2} r^4 \left(\frac{3}{a^5} + \frac{4}{b^5}\right) Y_4^0 + \left(\frac{35}{144}\right)^{1/2} \frac{r^4}{a^5}(Y_4^4 + Y_4^{-4})\right\} \tag{6-21}$$

Here we see that a second parameter enters the calculation, that being $\langle r^2 \rangle$.

Expressions for the crystal field states for some common geometries are given here in terms of radial parameters ρ_2 and ρ_4, where the latter have units of energy.

Square plane:

$$E(x^2 - y^2) = \frac{4}{7}\rho_2 + \frac{19}{21}\rho_4$$

$$E(z^2) = -\frac{4}{7}\rho_2 + \frac{3}{7}\rho_4$$

$$E(xz, yz) = -\frac{2}{7}\rho_2 - \frac{2}{7}\rho_4$$

$$E(xy) = \frac{4}{7}\rho_2 - \frac{16}{21}\rho_4$$

D_{2d} symmetry:

$$E(z^2) = 2Ds + 6Dt$$
$$E(x^2 - y^2) = -2Ds + Dt - Dr$$
$$E(xy) = -2Ds + Dt + Dr$$
$$E(xz, yz) = Ds - 4Dt$$

where

$$Ds = \tfrac{2}{7}(3\cos^2\alpha - 1)\rho_2$$
$$Dt = \tfrac{1}{14}(\tfrac{35}{3}\cos^4\alpha - 10\cos^2\alpha + 1)\rho_4$$
$$Dr = \tfrac{5}{6}(\sin^4\alpha)\rho_4$$
$$\alpha = \text{polar angle}$$

Trigonal bypyramid:

$$E(z^2) = \left(\frac{4}{7}\rho_2\right)^{ap} - \left(\frac{3}{7}\rho_2\right)^{eq} + \left(\frac{9}{28}\rho_4\right)^{eq} + \left(\frac{4}{7}\rho_4\right)^{ap}$$
$$E(xz, yz) = \left(\frac{2}{7}\rho_2\right)^{ap} - \left(\frac{3}{14}\rho_2\right)^{eq} - \left(\frac{8}{21}\rho_4\right)^{ap} - \left(\frac{3}{14}\rho_4\right)^{eq}$$
$$E(x^2 - y^2, xy) = \left(\frac{3}{7}\rho_2\right)^{eq} - \left(\frac{4}{7}\rho_2\right)^{ap} + \left(\frac{2}{21}\rho_4\right)^{ap} + \left(\frac{3}{56}\rho_4\right)^{eq}$$

(*eq* denotes equatorial bonds and *ap* apical bonds)

Tetragonally distorted octahedron:

$$E(z^2) = 6Dq - 2Ds - 6Dt$$
$$E(x^2 - y^2) = 6Dq + 2Ds - Dt$$
$$E(xy) = -4Dq + 2Ds - Dt$$
$$E(xz, yz) = -4Dq - Ds + 4Dt$$

where

$$Dq = (\tfrac{1}{6})\rho_4^{xy}$$
$$Ds = (\tfrac{2}{7})(\rho_2^{xy} - \rho_2^{z})$$
$$Dt = (\tfrac{2}{21})(\rho_4^{xy} - \rho_4^{z})$$

It is clear from these expressions that the number of parameters increases as the symmetry of the complex decreases.

Tetrahedral geometry

We shall now investigate the crystal field splitting in a ML_4 complex which belongs to point group T_d and which has a d^1 electronic configuration. The coordinate system to be used is shown in Figure 6.11.

The two-fold axis is chosen as the z-axis. Thus the ligands have the coordinates $(\pm a\sqrt{2/3}, 0, a\sqrt{1/3})$ and $(0, \pm a\sqrt{2/3}, -a\sqrt{1/3})$. Consider-

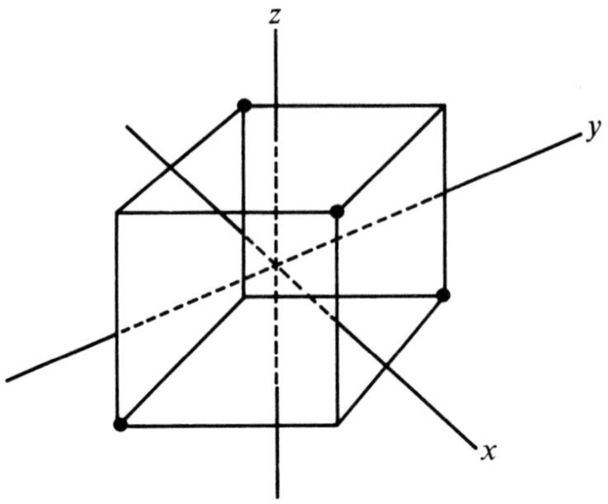

Figure 6.11
Coordinate System for the Tetrahedral Calculation

ing rotations only, we find

	E	$8C_3$	$3C_2$	
$\Gamma_{Y_2^m}$	5	-1	1	$= E + T_2$
$\Gamma_{Y_4^m}$	9	0	1	$= A_1 + \ldots$

and so that only fourth order spherical harmonics contribute to V_{T_d}. Choosing the C_2 axis as the quantization axis and noting that

$$S_4 \begin{bmatrix} x \\ y \\ z \end{bmatrix} = \begin{bmatrix} y \\ -x \\ -z \end{bmatrix}$$

we determine that

$$S_4 \begin{bmatrix} Y_4^4 \\ Y_4^3 \\ Y_4^2 \\ Y_4^1 \\ Y_4^0 \\ Y_4^{-1} \\ Y_4^{-2} \\ Y_4^{-3} \\ Y_4^{-4} \end{bmatrix} = \begin{bmatrix} Y_4^4 \\ -iY_4^3 \\ -Y_4^2 \\ iY_4^1 \\ Y_4^0 \\ -iY_4^{-1} \\ -Y_4^{-2} \\ iY_4^{-3} \\ Y_4^{-4} \end{bmatrix}$$

Therefore, V_{T_d} must be of the form $aY_4^0 + bY_4^4 + cY_4^{-4}$. Also, upon finding that $\sigma_d Y_4^4 = Y_4^{-4}$ and $\sigma_d Y_4^{-4} = Y_4^4$, we see that $V_{T_d} = eY_4^0 + f(Y_4^4 + Y_4^{-4})$.

Now we shall calculate the contributions to the crystal field potential energy operator. We have seen that

$$Y_4^{0*} = Y_4^0 = \sqrt{9/4\pi}(1/8)(35\cos^4\theta_i - 30\cos^2\theta_i + 3)$$

Therefore, the four ligands with these coordinates

point i	θ_i	$35\cos^4\theta_i - 30\cos^2\theta_i + 3$
$\sqrt{2/3}, 0, \sqrt{1/3}$	54°44′	$-28/9$
$0, \sqrt{2/3}, -\sqrt{1/3}$	125°16′	$-28/9$
$-\sqrt{2/3}, 0, \sqrt{1/3}$	54°44′	$-28/9$
$0, -\sqrt{2/3}, -\sqrt{1/3}$	125°16′	$-28/9$
		$-112/9$

contribute to the crystal field potential the term

$$\frac{4\pi}{9}\frac{Ze^2r^4}{a^5}\left(\frac{9}{4\pi}\right)^{1/2}\left(\frac{1}{8}\right)\left(\frac{-112}{9}\right)Y_4^0$$

For the $Y_4^{\pm4*} = Y_4^{\mp4}$ term we have

point i	θ_i	ϕ_i	$\sin^4\theta\, e^{\pm 4i\phi}$
$\sqrt{2/3}, 0, \sqrt{1/3}$	54°44′	0	4/9
$0, \sqrt{2/3}, -\sqrt{1/3}$	125°16′	$\pi/2$	4/9
$-\sqrt{2/3}, 0, \sqrt{1/3}$	54°44′	π	4/9
$0, -\sqrt{2/3}, -\sqrt{1/3}$	125°16′	$3\pi/2$	4/9
			16/9

and the contribution

$$\left(\frac{4\pi}{9}\right)^{1/2}\left(\frac{14}{9}\right)\left(\frac{5}{15}\right)^{1/2}\frac{Ze^2r^4}{a^5}(Y_4^4 + Y_4^{-4})$$

The completed expression for V_{T_d} is

$$V_{T_d} = \left(\frac{49}{18}\right)^{1/2}(2\pi)^{1/2}\left(\frac{-4}{9}\right)\frac{Ze^2r^4}{a^5}\left\{Y_4^0 - \left(\frac{5}{14}\right)^{1/2}(Y_4^4 + Y_4^{-4})\right\}$$

It is of interest to compare the tetrahedral crystal field potential with the octahedral one. However, we must modify V_{O_h} in Equation (6-15) before the comparison can be made since the coordinate systems used for the generation of the two potentials differ by 45° in ϕ. Since the angle ϕ affects only the $Y_4^{\pm 4}$ terms we find upon substitution that

$$V_{0_h} = \left(\frac{49}{18}\right)^{1/2} (2\pi)^{1/2} \frac{Ze^2r^4}{a^5}\left[Y_4^0 - \left(\frac{5}{14}\right)^{1/2} (Y_4^4 + Y_4^{-4})\right]$$

and that $V_{T_d} = -(4/9)V_{0_h}$.

This derivation suggests that the order of the energy levels for tetrahedral coordination will be reversed from that found for octahedral coordination and that the crystal field splitting, Δ_{T_d}, will be (4/9) Δ_{0_h}. In the discussion of the one electron correlation diagrams given in Figure 6.7 we had already anticipated the reversal in the ordering of the energy levels, and experimental results confirms that $\Delta_{T_d}/\Delta_{0_h}$ is rather close to the theoretically predicted ratio.

6-6 MANY-ELECTRON CORRELATION DIAGRAMS

Very useful diagrams that give an approximate ordering of the crystal field states for the configuration d^2 through d^8 can be constructed using ideas that were developed in Sections 3 and 4 of this chapter. To construct these diagrams, we plot on the left-hand side the energies of the free-ion Russell-Saunders terms, and on the right-hand side we have the relative energies of the octahedral crystal field configurations. For the d^2 ion, the configuration of lowest energy must be $(t_{2g})^2$. With the energy of this configuration taken as zero, $(t_{2g})(e_g)$ is at $+10\ Dq$ and the configuration $(e_g)^2$ is at $+20\ Dq$. These crystal field configurations will give rise to a number of energy states in the same manner as the free ion configurations, and there will be a one-to-one correspondence between these states and the states that will result from the splittings of the free ion Russell-Saunders terms. To illustrate the procedure we shall construct a correlation diagram for the d^2 ion in an octahedral crystal field. From Table 6.1 we see that the d^2 configuration gives rise to the R-S states ^3F, ^3P, ^1D, ^1G, ^1S, and from Table 6.2 we see that these R-S terms are split into the following crystal field states:

$$^3F\colon\ ^3A_{2g} + {}^3T_{1g} + {}^3T_{2g}$$
$$^3P\colon\ ^3T_{1g}$$
$$^1D\colon\ ^1E_g + {}^1T_{2g}$$
$$^1G\colon\ ^1A_{1g} + {}^1E_g + {}^1T_{1g} + {}^1T_{2g}$$
$$^1S\colon\ ^1A_{1g}$$

Of course, in an octahedral crystal field with a vanishingly small Dq, all the states arising from a particular Russell-Saunders term will have the same energy. This gives the left hand side of the correlation diagram.

To determine the crystal field states which arise from the configurations $(t_{2g})^2$, $(t_{2g})(e_g)$, and $(e_g)^2$, we must take the direct products of the representations, i.e., $t_{2g} \times t_{2g}$ for $(t_{2g})^2$, decompose these direct products to get the symmetries of the orbital wave functions, and then determine the multiplicities from the total degeneracy of the configurations. The states which arise will be either singlets or triplets because there are two electrons involved.

The direct products are:

$$t_{2g} \times t_{2g} = A_{1g} + E_g + T_{1g} + T_{2g}$$
$$e_g \times t_{2g} = T_{1g} + T_{2g}$$
$$e_g \times e_g = A_{1g} + A_{2g} + E_g$$

In order to determine the total degeneracies of the configurations we look at the options open for the quantum numbers. According to the following diagram for the configuration $(t_{2g})^2$

$m_s = 1/2$			
$m_s = -1/2$			
	t_{2g}^1	t_{2g}^2	t_{2g}^3

we see that one of the electrons can take on any of the six combinations of spin and orbital quantum numbers, while the second electron of the configuration is limited to one of the remaining five. Thus the total degeneracy is

$$\frac{6 \cdot 5}{2} = 15$$

where the factor of two in the denominator takes care of the indistinguishability of the electrons. This means that the multiplicities of the states

$${}^aA_{1g} + {}^bE_g + {}^cT_{1g} + {}^dT_{2g}$$

multiplied by the respective orbital degeneracies must equal 15. Thus

$$(1 \cdot a) + (2 \cdot b) + (3 \cdot c) + (3 \cdot d) = 15$$

where a, b, c, and d are equal to 1 or 3. Three solutions are possible:

$$a = b = c = 1 \qquad d = 3$$
$$a = b = d = 1 \qquad c = 3$$
$$a = b = 3 \qquad c = d = 1$$

For the configuration $(e_g)^2$:

	e_g^1	e_g^2
$m_s = 1/2$		
$m_s = -1/2$		

The total degeneracy is $(4 \cdot 3)/2 = 6$, and for ${}^aA_{1g} + {}^bA_{2g} + {}^cE_g$, then

$$(1 \cdot a) + (1 \cdot b) + (2 \cdot c) = 6$$

Two solutions are possible. These are:

$$a = c = 1 \qquad b = 3$$
$$b = c = 1 \qquad a = 3$$

For the configuration $(t_{2g})(e_g)$, the first electron can have any of the six t_{2g} sets of quantum numbers, and the second any of the four e_g sets. This gives the total degeneracy equal to 24, and the states are

$$^1T_{1g} \quad ^3T_{1g} \quad ^1T_{2g} \quad ^3T_{2g}$$

The correlation diagram shown in Figure 6.12 results. As noted earlier there must be a one-to-one correspondence of states on the left arising from the splitting of the Russell-Saunders terms with those derived from the crystal field configurations on the right. This allows the multiplicities to be assigned. We see that there are two $^3T_{1g}$ states on the left, therefore there must be two $^3T_{1g}$ states on the right. We have one $^3T_{1g}$ from the $(t_{2g})(e_g)$ configuration, and the second must come from $(t_{2g})^2$ (there being no T_{1g} in $e_g \times e_g$). This means that the states $^1A_{1g}$, ^1E_g, $^1T_{2g}$, and $^3T_{1g}$ arise from $(t_{2g})^2$. Also, there is a $^3A_{2g}$ on the left, and to get a $^3A_{2g}$ on the right we pick the solution for $(e_g)^2$ of $^1A_{1g}$, ^1E_g, $^3A_{2g}$.

In drawing the tie lines in Figure 6.12 to correlate the crystal field states with the free ion states, the *noncrossing rule* was obeyed. The *noncrossing rule* says that tie lines representing states with the same symmetry and multiplicity can never cross.

Exercise 6-3

Construct a correlation diagram for a d^2 ion in a complex ML_5 with C_{4v} symmetry. Assume the *d*-orbitals have the following energies: $d_{x^2-y^2}$ (9.14 *Dq*), d_{z^2}(0.86 *Dq*), d_{xy}(−.86 *Dq*), d_{xz}, d_{yz}(−4.57 *Dq*). The completed diagram is given in Appendix 6 B.

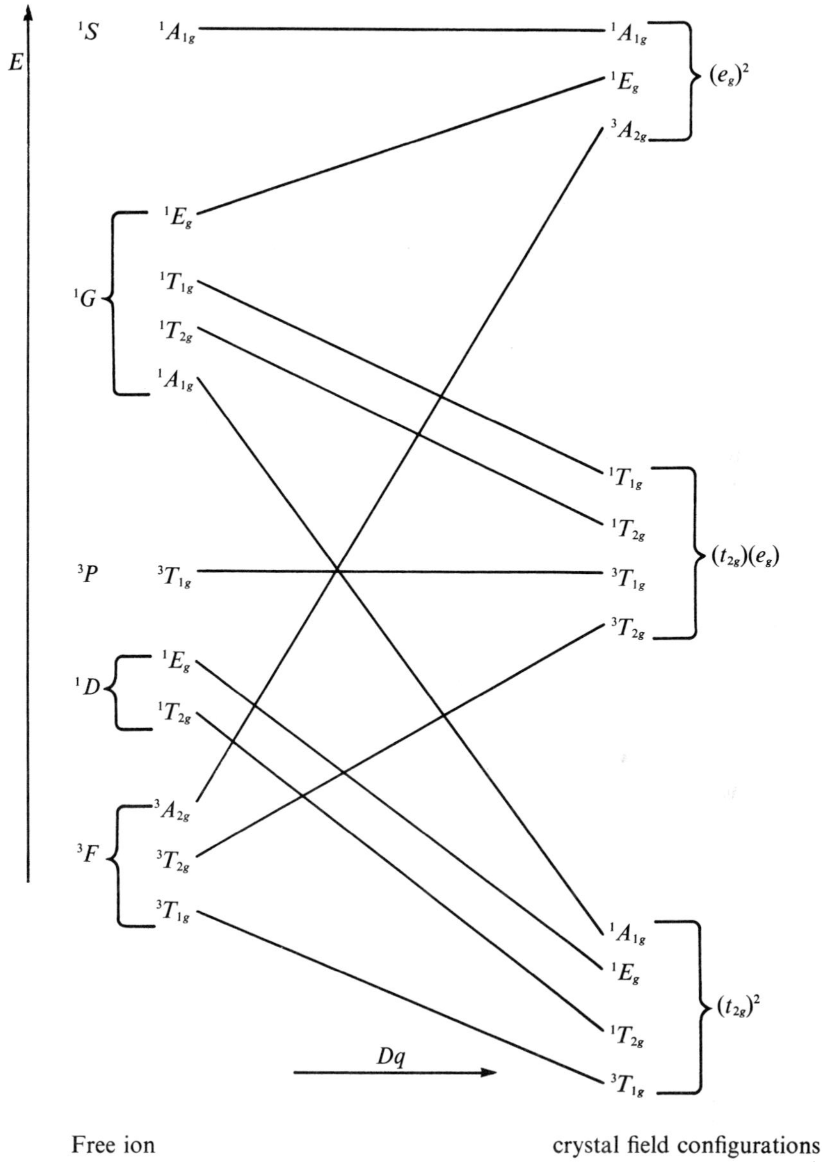

Figure 6.12

Correlation Diagram for an Ion with d^2 Configuration in an Octahedral Crystal Field

The correlation diagram for an ion with a d^2 electronic configuration in a tetrahedral environment is constructed in the same manner. The splitting of the Russell Saunders states is found to be

$$
\begin{aligned}
{}^1S &: \ {}^1A_1 \\
{}^1G &: \ {}^1A_1 + {}^1E + {}^1T_1 + {}^1T_2 \\
{}^3P &: \ {}^3T_1 \\
{}^1D &: \ {}^3T_1 \\
{}^3F &: \ {}^3A_2 + {}^3T_1 + {}^3T_2
\end{aligned}
$$

This gives the required information for the left side of the diagram. The crystal field configurations in order of increasing energy are $(e)^2$, $(e)(t_2)$, and $(t_2)^2$. Forming the direct products and reducing the representations, we find

$$
\begin{aligned}
e \times e &= A_1 + A_2 + E \\
t_2 \times e &= T_1 + T_2 \\
t_2 \times t_2 &= A_1 + E + T_1 + T_2
\end{aligned}
$$

The crystal field states arising from the configurations are arranged on the right of the correlation chart and tie lines are drawn. It is important to note the similarity between the $d^2 - O_h$ and $d^2 - T_d$ correlation charts. Essentially the only differences in the two diagrams is that there are g subscripts on the O_h states, and the configurations on the right side of the chart are inverted.

There is a useful relationship between correlation charts for the d^n and d^{10-n} configurations. For example, in a tetrahedral complex of nickel(II) the following electronic configuration is found:

$\uparrow\downarrow$ $\uparrow$ $\uparrow$ t_2

$\uparrow\downarrow$ $\uparrow\downarrow$ e

This may be considered as two holes in the t_2 levels. Also, as shown in Table 6.1, the same Russell-Saunders terms arise for both d^2 and d^8 configurations. Thus, the correlation diagram derived for the $d^2 - O_h$ case applies equally well to the $d^8 - T_d$ case. In general the $d^n - O_h$ correlation chart is identical to the $d^{10-n} - T_d$ chart. Therefore, only four correlation charts need be constructed to describe the crystal field states in the fourteen possibilities for d^2 through d^8 configurations and O_h or T_d geometries. Some especially useful correlation charts will be discussed later in this chapter.

6-7 CRYSTAL FIELD CALCULATION FOR TWO ELECTRONS

The crystal field calculation in the weak field scheme for two (or more) d-electrons proceeds in essentially the same manner as that described for one d-electron. The seven orbital wave functions for the ^3F ground state of the free ion are used in the perturbation calculation with the octahedral crystal field potential. The crystal field potential energy operator is simply written as

$$V_{O_h}^{d_2} = V_{O_h}^{1} + V_{O_h}^{2}$$

where the superscripts designate the two electrons. Recall that the summation in Equation (6-9) was over the number of electrons too.

The seven orbital wave functions are constructed from the information found in Appendix 6A. The M_L values for an F term run from $+3$ to -3, and from the chart for the d^2 configuration we find the products of one-electron functions shown in Figure 6.13, where d_2, d_1, d_0, d_{-1}, d_{-2} specify the hydrogen-like d-wave functions as designated by the m_l quantum number used as a subscript.

M_L	
3	(d_2, d_1)
2	(d_2, d_0)
1	$(d_2, d_{-1})(d_1, d_0)$
0	$(d_2, d_{-1})(d_1, d_{-1})$
-1	$(d_0, d_{-2})(d_0, d_{-1})$
-2	(d_0, d_{-2})
-3	(d_{-1}, d_{-2})

Figure 6.13
Portion of Micro-State Chart for ^3F State

Representing our two-electron wave functions as $\Psi(L, M_L)$, it is readily seen that four of the functions are

$$\Psi(3, 3) = d_2 d_1$$

$$\Psi(3, 2) = d_2 d_0$$

$$\Psi(3, -2) = d_0 d_{-2}$$

$$\Psi(3, -3) = d_{-1} d_{-2}$$

since these are the only products of one electron wave functions which will give the appropriate M_L value. Also, the one electron d_i-functions are presumably normalized; so the resultant products are normalized. However, a problem exists with respect to the $\Psi(3, 1)$, $\Psi(3, 0)$, $\Psi(3, -1)$ functions since there are two products of one-electron functions which have the requisite M_L values. Since we have no good reason for choosing one of these functions over the other, we must take a linear combination of both. To do this we use the raising and lowering, or ladder, operators. For this calculation we will use the lowering operator L- which is defined as

$$L\text{-}\Psi(L, M_L) = \sqrt{(L - M_L + 1)(L + M_L)}\,\Psi(L, M_L - 1) \quad \textbf{(6-22)}$$

Starting with

$$\Psi(3, 2) = d_2 d_0$$

we apply L- (to both sides) and obtain

$$\sqrt{(3 - 2 + 1)(3 + 2)}\,\Psi(3, 1) = \sqrt{(2 - 2 + 1)(2 + 2)}\,(d_1 d_0) + \sqrt{(2 - 0 + 1)(2 + 0)}\,(d_2 d_{-1})$$

The function $\Psi(3, 1)$ results:

$$\sqrt{10}\,\Psi(3, 1) = \sqrt{4}\,(d_1 d_0) + \sqrt{6}\,(d_2 d_{-1})$$

$$\Psi(3, 1) = \sqrt{2/5}\,(d_1 d_0) + \sqrt{3/5}\,(d_2 d_{-1}) \quad \textbf{(6-23)}$$

The linear combination for $\Psi(3, 0)$ is generated by application of L- on $\Psi(3, 1)$:

$$L\text{-}\Psi(3, 1) = \sqrt{12}\,\Psi(3, 0)$$
$$= \sqrt{12/5}\,d_1 d_{-1} + \sqrt{12/5}\,d_1 d_{-1} + \sqrt{12/5}\,d_2 d_{-2}$$

and

$$\Psi(3, 0) = \sqrt{4/5}\,d_1 d_{-1} + \sqrt{1/5}\,d_2 d_{-2} \quad \textbf{(6-24)}$$

By the same methods we find

$$\Psi(3, -1) = \sqrt{2/5}\,d_0 d_{-1} + \sqrt{3/5}\,d_1 d_{-2} \quad \textbf{(6-25)}$$

The seven orbital wave functions for the F state need to be adjusted to meet the requirements of the Pauli principle, which stipulates that there must be a change in the sign of the wave functions if the coordinates of any two electrons are interchanged. The general expression for wave functions which obey the Pauli principle is given in Equation (6-26).

$$|\Psi\rangle = \frac{1}{\sqrt{N!}}\begin{vmatrix} |n_1 l_1 m_1\rangle^1 & |n_1 l_1 m_1\rangle^2 & \cdots & |n_1 l_1 m_1\rangle^N \\ |n_2 l_2 m_2\rangle^1 & |n_2 l_2 m_2\rangle^2 & \cdots & |n_2 l_2 m_2\rangle^N \\ \cdots & \cdots & \cdots & \cdots \\ |n_N l_N m_N\rangle^1 & |n_N l_N m_N\rangle^2 & \cdots & |n_N l_N m_N\rangle^N \end{vmatrix} \qquad \textbf{(6-26)}$$

where the superscript j on the ket $|n_i l_i m_i>^j$ designates the electron whose coordinates are to appear in the orbital designated by the indicated quantum numbers i.

We are now ready to calculate the matrix elements of the energy determinantal equation.

$$\begin{array}{c|ccccccc|} & 3 & 2 & 1 & 0 & -1 & -2 & -3 \\ 3 & & & & & & & \\ 2 & & & & & & & \\ 1 & & & & & & & \\ 0 & & & & & & & \\ -1 & & & & & & & \\ -2 & & & & & & & \\ -3 & & & & & & & \end{array} = 0$$

We will show the calculation for one element and give the results for the remainder. For the first element $\langle\psi(3)|\, V^1_{O_h} + V^2_{O_h}\,|\psi(3)\rangle$ we have

$$\left\langle \frac{1}{\sqrt{2}}\begin{vmatrix} d_2^1 & d_2^2 \\ d_1^1 & d_1^2 \end{vmatrix} (V^1_{O_h} + V^2_{O_h}) \begin{vmatrix} d_2^1 & d_2^2 \\ d_1^1 & d_1^2 \end{vmatrix} \frac{1}{\sqrt{2}} \right\rangle$$

$$\begin{aligned} &= \tfrac{1}{2}\langle (d_2^1 d_1^2 - d_1^1 d_2^2)|\, V^1_{O_h} + V^2_{O_h}\,|(d_2^1 d_1^2 - d_1^1 d_2^2)\rangle \\ &= \tfrac{1}{2}\{\langle d_2^1 d_1^2|V^1_{O_h}|d_2^1 d_1^2\rangle + \langle d_2^1 d_1^2|V^2_{O_h}|d_2^1 d_1^2\rangle \\ &\quad -\langle d_2^1 d_1^2|V^1_{O_h}|d_1^1 d_2^2\rangle - \langle d_2^1 d_1^2|V^2_{O_h}|d_1^1 d_2^2\rangle \\ &\quad -\langle d_1^1 d_2^2|V^1_{O_h}|d_2^1 d_1^2\rangle - \langle d_1^1 d_2^2|V^2_{O_h}|d_2^1 d_1^2\rangle \\ &\quad +\langle d_1^1 d_2^2|V^1_{O_h}|d_1^1 d_2^2\rangle + \langle d_1^1 d_2^2|V^2_{O_h}|d_1^1 d_2^2\rangle\} \end{aligned}$$

The crystal field potential operates on only one electron; so for example,

$$\begin{aligned} \langle d_2^1 d_1^2|V^1_{O_h}|d_2^1 d_1^2\rangle &= \langle d_2^1|V^1_{O_h}|d_2^1\rangle\langle d_1^2|d_1^2\rangle \\ &= \langle d_2^1|V^1_{O_h}|d_2^1\rangle \end{aligned}$$

since $\langle d_1^2 | d_1^2 \rangle = 1$ by normalization. We have evaluated integrals like $\langle d_2^1 | V_{O_h}^1 | d_2^1 \rangle$ in Section 6-5, using Eq. (6-19). Using those results, we get

$$\begin{aligned} \langle \psi(3) | V_{O_h}^{d_2} | \psi(3) \rangle &= \tfrac{1}{2}\{Dq - 4Dq + 0 + 0 + 0 + 0 - 4Dq + Dq\} \\ &= -3Dq \end{aligned}$$

The completed secular determinantal equation is shown in Figure 6.14. The determinant may be factored to yield the following smaller determinantal equations:

$$\begin{array}{cc} 3 & -1 \end{array}$$
$$A \begin{vmatrix} -3Dq - E & \sqrt{15}\,Dq \\ \sqrt{15}\,Dq & -Dq - E \end{vmatrix} = 0$$

$$\begin{array}{cc} 2 & -2 \end{array}$$
$$B \begin{vmatrix} 7Dq - E & 5Dq \\ 5Dq & 7Dq - E \end{vmatrix} = 0$$

$$\begin{array}{cc} 1 & -3 \end{array}$$
$$C \begin{vmatrix} -Dq - E & \sqrt{15}\,Dq \\ \sqrt{15}\,Dq & -3Dq - E \end{vmatrix} = 0$$

$$D \; | -6Dq - E | = 0$$

where equations A and C are identical. The roots are:

$$\begin{aligned} (A, C) \quad 3Dq^2 + 4DqE + E^2 - 15Dq &= 0 \\ E^2 + 4DqE - 12Dq^2 &= 0 \\ (E + 6Dq)(E - 2Dq) &= 0 \\ E &= -6Dq \\ E &= 2Dq \\ (B) \qquad E &= 12Dq \\ E &= 2Dq \\ (D) \qquad E &= -6Dq \end{aligned}$$

From Table 6.2 we know that the F term breaks up into two triply degenerate levels and one nondegenerate level. The results of our calculation show the triply degenerate levels to lie at $-6Dq$ and $2Dq$ while the nondegenerate level is at 12 Dq.

$$
\begin{vmatrix}
 & 3 & 2 & 1 & 0 & -1 & -2 & -3 \\
3 & -3Dq - E & 0 & 0 & 0 & \sqrt{15}Dq & 0 & 0 \\
2 & 0 & 7Dq - E & 0 & 0 & 0 & 5Dq & 0 \\
1 & 0 & 0 & -Dq - E & 0 & 0 & 0 & \sqrt{15}Dq \\
0 & 0 & 0 & 0 & -6Dq - E & 0 & 0 & 0 \\
-1 & \sqrt{15}Dq & 0 & 0 & 0 & -Dq - E & 0 & 0 \\
-2 & 0 & 5Dq & 0 & 0 & 0 & 7Dq - E & 0 \\
-3 & 0 & 0 & \sqrt{15}Dq & 0 & 0 & 0 & -3Dq - E
\end{vmatrix} = 0
$$

Figure 6.14
Secular Determinantal Equation for Crystal Field Calculation for Two Electrons in O_h *Field*

By substitution of the derived energies into the secular equations the crystal field corrected wave functions are found to be

$$
\begin{array}{ll}
E = 12\,Dq & [\Psi(3, 2) - \Psi(3, -2)]/\sqrt{2} \\
E = 2\,Dq & \left\{\begin{array}{l} \sqrt{5/8}\,\Psi(3, 1) - \sqrt{3/8}\,\Psi(3, -3) \\ \sqrt{5/8}\,\Psi(3, -1) - \sqrt{3/8}\,\Psi(3, 3) \\ [\Psi(3, 2) + \Psi(3, -2)]/\sqrt{2} \end{array}\right. \\
E = -6\,Dq & \left\{\begin{array}{l} \Psi(3, 0) \\ \sqrt{3/8}\,\Psi(3, 1) + \sqrt{5/8}\,\Psi(3, -3) \\ \sqrt{3/8}\,\Psi(3, -1) + \sqrt{5/8}\,\Psi(3, 3) \end{array}\right.
\end{array}
\qquad \textbf{(6-27)}
$$

Although we could construct the representations to which these functions belong by the usual means, we can arrive at the symmetry labels with little work by noting that there is only one nondegenerate level, and then it is immediately known that the ${}^3A_{2g}$ is at 12 Dq. Also in the last section (in Figure 6.12) we found that the ground state of the d^2 configuration in an O_h field was the ${}^3T_{1g}$, so the other triply degenerate level at 2 Dq must be the ${}^3T_{2g}$ state.

By reference to Table 6.2 we see that a ${}^3T_{1g}$ state also arises from the ${}^3P$ level. Since this state has the same symmetry as the ground state, the two may interact in such a way that the ground state is stabilized and ${}^3T_{1g}(P)$ is destabilized.

To calculate the magnitude of this configuration interaction we need the functions for $\Psi(1, 1)$. For the ${}^3T_{1g}(F)$ we know from Equations (6.23) through (6.25) that

$$
\begin{aligned}
\Psi(3, 1) &= \sqrt{3/5}\, d_2 d_{-1} + \sqrt{2/5}\, d_1 d_0 \\
\Psi(3, 0) &= \sqrt{4/5}\, d_1 d_{-1} + \sqrt{1/5}\, d_2 d_{-2} \\
\Psi(3, -1) &= \sqrt{2/5}\, d_0 d_{-1} + \sqrt{3/5}\, d_1 d_{-2}
\end{aligned}
$$

and since the ${}^3T_{1g}(P)$ functions must be orthogonal we can write

$$
\begin{aligned}
\Psi(1, 1) &= \sqrt{2/5}\, d_2 d_{-1} - \sqrt{3/5}\, d_1 d_0 \\
\Psi(1, 0) &= \sqrt{4/5}\, d_2 d_{-2} - \sqrt{1/5}\, d_1 d_{-1} \\
\Psi(1, -1) &= \sqrt{2/5}\, d_1 d_{-2} - \sqrt{3/5}\, d_0 d_{-1}
\end{aligned}
$$

The 6×6 determinant obtained factors into three identical 2×2 determinants of the form

$$
\begin{array}{l} T_{1g}(F) \\ T_{1g}(P) \end{array}
\begin{vmatrix} -6Dq - E & 4Dq \\ 4Dq & p - E \end{vmatrix} = 0
$$

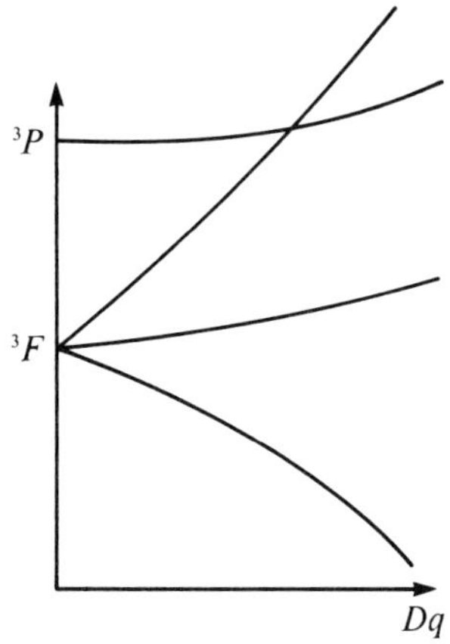

Figure 6.15
Energy Level Diagram for ^3F and ^3P Terms of d^2 in an O_h Field

where p is the $^3F - {}^3P$ term energy separation, and the off diagonal element is calculated in the usual manner.

The results of the calculation are shown in Figure 6.15.

We now can define a difference between crystal field theory and ligand field theory: If p is taken from the spectrum of the free ion and kept constant, then pure crystal field theory is being used. However, if p is taken as a variable, and permitted to decrease from the free ion value, then ligand field theory is used. This is the manner in which the theory has been modified to account for covalency in the metal-ligand bonds. If the metal electrons are delocalized to some extent onto the ligands, then interelectronic repulsions are expected to decrease and the $^3F - {}^3P$ term separation should decrease.

6-8 SPIN-ORBIT COUPLING

For an understanding of many of the properties of transition metal compounds it is necessary to consider the effects of spin-orbit coupling. This magnetic interaction takes place between the spin magnetic moment and the orbital magnetic moment of the electrons. The term that is added to the Hamiltonian is

$$H_{so} = \alpha \sum_i \zeta_i \vec{l_i} \cdot \vec{s_i}$$

where α is the fine structure constant and ζ is a function of the radial coordinate. If interactions between different Russell-Saunders states are not large, then we may write

$$H_{so} = \lambda_{LS}(\vec{L}\cdot\vec{S}) \tag{6-28}$$

where for a shell less than a half filled

$$\lambda_{LS} = \frac{\zeta}{2S}$$

but for a shell more than half filled

$$\lambda_{LS} = -\frac{\zeta}{2S}$$

and

$$\begin{aligned}\vec{L}\cdot\vec{S}\,|\,\Psi(M_L, M_S)> &= (L_zS_z + \tfrac{1}{2}(L_x + iL_y)(S_x - iS_y) \\ &+ \tfrac{1}{2}(L_x - iL_y)(S_x + iS_y))\,|\,\Psi(M_L, M_S)> = M_LM_S\,|\,\Psi(M_L, M_S)> \\ &+ \tfrac{1}{2}\sqrt{(L-M_L)(L+M_L+1)(S+M_S)(S-M_S+1)}\,|\,\Psi(M_L+1, M_S-1)> \\ &+ \tfrac{1}{2}\sqrt{(L+M_L)(L-M_L+1)(S-M_S)(S+M_S+1)}\,|\,\Psi(M_L-1, M_S+1)>\end{aligned} \tag{6-29}$$

Here, $\vec{L}\cdot\vec{S}$ has been expanded in terms of the raising and lowering operators, which were introduced earlier.

The spin-orbit interaction becomes very important in the case of heavier atoms, and in many cases the states are characterized by the total angular momentum $\vec{J}$. Since $\vec{J} = \vec{L} + \vec{S}$ then $\vec{J}$ can take on both integer and half integer values, which causes a problem to arise. Up to now we have defined a rotation by 360° to be equivalent to the identity operation and that

$$\chi(\alpha) = \chi(\alpha + 2\pi)$$

Consider a case where J is a half integer. Application of Eq. (6-7) yields

$$\begin{aligned}\chi(\alpha + 2\pi) &= \frac{\sin (J + 1/2)(\alpha + 2\pi)}{\sin (\alpha + 2\pi)/2} \\ &= \frac{\sin [(J + 1/2)\alpha + n2\pi]}{\sin (\alpha/2 + \pi)} \\ &= \frac{\sin (J + 1/2)\alpha}{-\sin \alpha/2} \\ &= -\chi(\alpha)\end{aligned}$$

In order to get out of this difficulty Bethe invented *double groups.* In a double group the rotation by 2π is treated as a symmetry operation but not as the identity operation. To construct the double group the products of the new operation (R) with all the symmetry operations in the original group are taken.

The classes in the double rotation groups can be shown to be

(1) E

(2) R

(3) C_2 and C_2R

(4) C_n and $C_n^{n-1}R$

(5) C_n^m and $C_n^{n-m}R$

and $\chi(E) = 2J + 1$ while $\chi(R) = 2J + 1$ if J is an integer or $\chi(R) = -(2J + 1)$ if J is a half integer.

Character tables for some useful double groups are given in Appendix 6C.

We shall now consider spin-orbit coupling in the complex ion $CuCl_4^{2-}$. Spin-orbit coupling is not large here and the procedure is to first find the effect of the crystal field on the free ion term and then consider the spin-orbit interaction as a perturbation. In Cs_2CuCl_4 the ion occupies a site in the crystal lattice that has symmetry close to D_{2d} with the major distortion from a tetrahedron being a compression along the z-axis as shown in Figure 6.16. The expected energy level diagram based on simple electrostatic considerations is also given in Figure 6.16. In Figure 6.16 we have given the now familiar symmetry labels ($^2A_1$, $^2B_1$, ^2E, $^2B_2$) for the states as

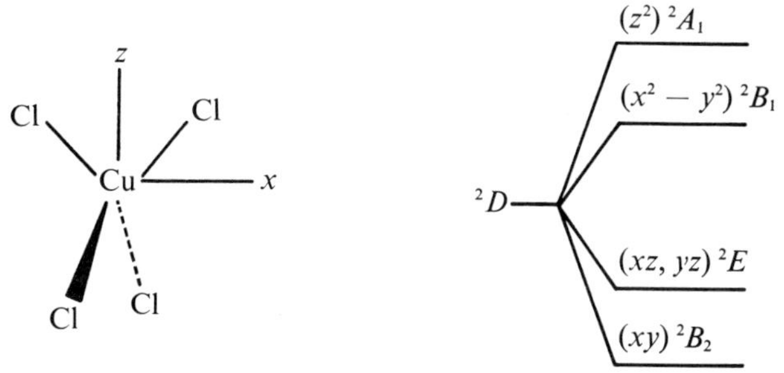

Figure 6.16
Structure and Energy Level Diagram for $CuCl_4^{2-}$

determined by the spatial properties of the orbitals. For finite spin-orbit coupling the total wave function for the states must be a product of the orbital function times a spin function. To determine the representation to which this product function belongs we must take the direct product of the representation for the orbital part times the representation to which the spin part belongs.

For the d^9 copper(II) ion $S = 1/2$ and the spin function transforms as Γ_6 in the double group D_{2d}. The required direct products and resultant spin-orbit representations are

$$\begin{aligned} \Gamma_6 \times A_1 &= \Gamma_6 \\ \Gamma_6 \times B_1 &= \Gamma_7 \\ \Gamma_6 \times E &= \Gamma_6 + \Gamma_7 \\ \Gamma_6 \times B_2 &= \Gamma_7 \end{aligned}$$

Calculation for d^1 Configuration in O_h

For the calculation of spin-orbit coupling for a metal ion with a d^1 electronic configuration in an octahedral crystal field, we will consider only the interactions within the ${}^2T_{2g}$ ground state. Although a neglect of the ${}^2E_g$ state can not be justified for the description of the properties observed in many experiments, the abbreviation will suffice for the purposes of our discussions.

The orbital parts of the crystal field corrected wave functions for the ${}^2T_{2g}$ level were found in Section 6.4 to be

$$\begin{aligned} &(d_2 - d_{-2})/\sqrt{2} \\ &d_1 \\ &d_{-1} \end{aligned}$$

The functions including spin become

$$\begin{aligned} \phi_1 &= (1/\sqrt{2})(d_2 - d_{-2})\alpha \\ \phi_2 &= (1/\sqrt{2})(d_2 - d_{-2})\beta \\ \phi_3 &= d_1\alpha \\ \phi_4 &= d_1\beta \\ \phi_5 &= d_{-1}\alpha \\ \phi_6 &= d_{-1}\beta \end{aligned} \qquad \textbf{(6-30)}$$

where α indicates $m_s = +1/2$ and β indicates $m_s = -1/2$.

It was noted above that the spin-orbit coupling operator

$$H_{so} = \lambda \vec{L} \cdot \vec{S}$$

can be expanded in terms of the raising and lowering operators (Equation (6-29)). In degenerate perturbation theory we must use the matrix elements of the 6×6 secular determinant which arise with the basis set of functions (Equation (6-30)). To facilitate this calculation we will first operate on Equation (6-30) with the operator $\vec{L}\cdot\vec{S}$. For example,

$$\begin{aligned}
\vec{L}\cdot\vec{S}|\phi_1\rangle &= \frac{1}{2}(L_x + iL_y)(S_x - iS_y)\left|\left(\frac{1}{\sqrt{2}}\right)(d_2 - d_{-2})\alpha\right\rangle \\
&= \frac{1}{2\sqrt{2}}(L_x + iL_y)|(d_2 - d_{-2})\beta\rangle \\
&= -\left(\frac{1}{\sqrt{2}}\right)|d_{-1}\beta\rangle \\
&= -\left(\frac{1}{\sqrt{2}}\right)|\phi_6\rangle
\end{aligned}$$

The complete results are

$$\begin{aligned}
\vec{L}\cdot\vec{S}|\phi_2\rangle &= \left(\frac{1}{\sqrt{2}}\right)|\phi_3\rangle \\
\vec{L}\cdot\vec{S}|\phi_3\rangle &= |d_2\beta\rangle + \tfrac{1}{2}|\phi_3\rangle \\
\vec{L}\cdot\vec{S}|\phi_4\rangle &= -(\tfrac{1}{2})|\phi_4\rangle \\
\vec{L}\cdot\vec{S}|\phi_5\rangle &= -(\tfrac{1}{2})|\phi_5\rangle \\
\vec{L}\cdot\vec{S}|\phi_6\rangle &= |d_{-2}\alpha\rangle + \tfrac{1}{2}|\phi_6\rangle
\end{aligned}$$

With these results we can write the secular determinant shown in Figure 6.17. This 6×6 determinant factors to yield the determinantal equations

$$A: \begin{vmatrix} -E & -\lambda/\sqrt{2} \\ -\lambda/\sqrt{2} & \lambda/2 - E \end{vmatrix} = 0$$

$$B: \begin{vmatrix} -E & \lambda/\sqrt{2} \\ \lambda/\sqrt{2} & \lambda/2 - E \end{vmatrix} = 0$$

$$C: |-\lambda/2 - E| = 0$$

$$D: |-\lambda/2 - E| = 0$$

These give rise to the solutions:

$E = -\lambda/2$	twice (from C and D)
	once from A
	once from B
$E = \lambda$	once from A
	once from B

$$\begin{array}{cccccc} \phi_1 & \phi_2 & \phi_3 & \phi_4 & \phi_5 & \phi_6 \end{array}$$

$$\begin{vmatrix} -E & 0 & 0 & 0 & 0 & -\lambda/\sqrt{2} \\ 0 & -E & \lambda/\sqrt{2} & 0 & 0 & 0 \\ 0 & \lambda/\sqrt{2} & \lambda/2 - E & 0 & 0 & 0 \\ 0 & 0 & 0 & -\lambda/2 - E & 0 & 0 \\ 0 & 0 & 0 & 0 & -\lambda/2 - E & 0 \\ -\lambda/\sqrt{2} & 0 & 0 & 0 & 0 & \lambda/2 - E \end{vmatrix} = 0$$

Figure 6.17
Secular Determinantal Equation for Spin-Orbit Interaction in $^2T_{2g}$ State

Thus, under the influence of spin-orbit coupling, the six-fold degenerate $^2T_{2g}$ state is split into a four-fold degenerate level at $-\lambda/2$ and a doubly degenerate level at λ. Substitution of these energies into the secular equations yields the wave functions:

$$\begin{aligned} \lambda:\ \Psi_1 &= (\phi_2 - \sqrt{2}\,\phi_3)/\sqrt{3} \\ \Psi_2 &= (\phi_1 - \sqrt{2}\,\phi_6)/\sqrt{3} \\ -\tfrac{1}{2}\lambda:\ \Psi_3 &= (\sqrt{2}\,\phi_2 - \phi_3)/\sqrt{3} \\ \Psi_4 &= (\sqrt{2}\,\phi_1 + \phi_6)/\sqrt{3} \\ \Psi_5 &= \phi_4 \\ \Psi_6 &= \phi_5 \end{aligned} \qquad \textbf{(6-31)}$$

The direct product of $\Gamma(S = 1/2)$ with the T_2 irreducible representation yields $\Gamma_7 + \Gamma_8$ where Γ_7 is the doubly degenerate representation and Γ_8 is the four-fold representation.

6-9 SPECTROSCOPY AND CRYSTAL FIELD THEORY

Spectra of Diatomic Molecules

Some of the features essential for the understanding of electronic spectra can most readily be illustrated by examining the ground state and an excited state of a diatomic molecule. In Figure 6.18 we show the potential energy curves which describe these two states. The vibrational energy levels are indicated as the horizontal lines, and these are referred to as the $v_0, v_1, v_2 \ldots v_n$ vibrational levels. For the purposes of this discussion rotational energy levels will be ignored.

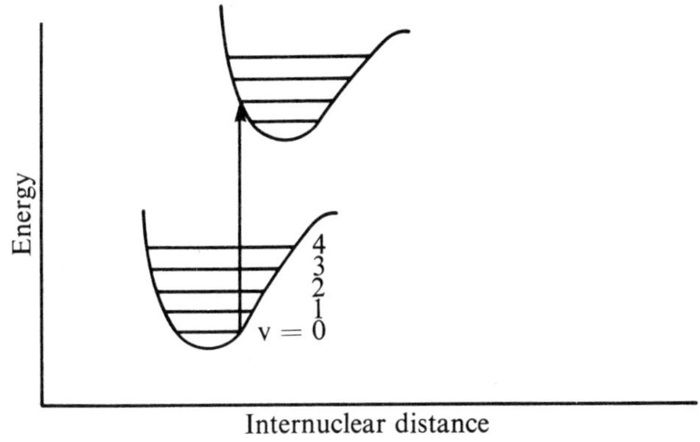

Figure 6.18
Potential Energy Curves

Transitions occur from the populated vibrational levels of the ground state to some vibrational level in the excited state. There is no selection rule that governs the change in the vibrational level during an electronic transition. However, according to the Franck-Condon principle, the atoms in the molecule do not change position during an electronic transition. This follows since electronic transitions can occur in a very short interval of time. Therefore, we represent the electronic transition in Figure 6.18 as a vertical line signifying that there has been no change in the internuclear distance.

Suppose that the transition in Figure 6.18 corresponded to the excitation of an electron from a bonding to an antibonding orbital. Obviously, the equilibrium internuclear distance of the two states will differ, and in some instances the differences may be great enough to preclude transitions from $v = 0$ in the ground state to the $v' = 0$ or $v' = 1$ level in the excited state.

Spectra of Polyatomic Molecules

Because of the greater number of degrees of freedom, the potential energies of polyatomic molecules must be represented as surfaces. Even so the basic ideas mentioned above apply to these systems. In addition it is important to recognize that the geometry of the excited state may not be the same as that of the ground state. For example, CO_2 has a linear structure in the ground state, but there are good reasons to believe that the structure is nonlinear in one or more of the excited states.

Selection Rules

The intensities of electronic absorption bands are proportional to integrals of the type

$$\int \Psi_{\text{e.s.}} \mu \Psi_{\text{g.s.}} \, d\tau \qquad \textbf{(6-32)}$$

where $\Psi_{\text{e.s.}}$ and $\Psi_{\text{g.s.}}$ are the wave functions of the excited and ground states, respectively, and μ is the dipole moment operator. The integrand will be nonzero only if the direct product of the representations of the wave functions and the operator yields the totally symmetric irreducible representation of the point group. The components of the electric dipole moment operator transform as x, y, and z. Transitions between electronic states may be allowed by other processes such as magnetic dipolar and electric quadrapolar processes, but since all spin-allowed transitions have been found to be electric dipolar in origin, we will discuss selection rules only for the electric dipole process.

If the molecule under consideration has a center of symmetry, then the integrand will always be zero because the electric dipole moment operator will transform as an odd irreducible representation, that is as u, and the wave functions for the ground and excited states must transform as even (g) irreducible representations. This follows since $g \times u \times g = u$. However, we know that electronic transitions occur for octahedral molecules in which there is a center of symmetry. This is an example of the failure of the selection rules due to the break-down of the Born-Oppenheimer approximation. In other words, the vibrational motions couple with the electronic motions, and the transitions may become allowed by a vibronic mechanism. In this case, the integrals are of the form

$$\int \Psi^{\text{excited}}_{\text{electronic}} \Psi^{\text{excited}}_{\text{vibrational}} \mu \Psi^{\text{ground}}_{\text{electronic}} \, d\tau \qquad \textbf{(6-33)}$$

The ground vibrational state always transforms as the totally symmetric irreducible representation and need not be considered. To find out whether a transition is vibronically allowed, first determine the symmetry of the integrals,

$$\int \Psi^{\text{excited}}_{\text{electronic}} \mu \Psi^{\text{ground}}_{\text{electronic}} \, d\tau$$

then check for the symmetries of the normal modes of vibration. If a normal mode of vibration forms the basis for a representation which transforms the same way as

$$\int \Psi^{\text{excited}}_{\text{electronic}} \mu \Psi^{\text{ground}}_{\text{electronic}} d\tau,$$

then these may couple and the transition becomes allowed. Recall that $\Gamma_i \times \Gamma_i$ always gives a representation which contains the totally symmetric irreducible representation.

These orbital selection rules are very useful in the analysis of the electronic spectrum, as will be demonstrated later. The real value of these selection rules is reached only when spectra are recorded using oriented single crystals and polarized light.

In describing these selection rules we have made no mention of the multiplicity of the states. Since the dipole moment operator does not affect the spin in the approximation described above, transitions between states with different multiplicities are forbidden due to the orthogonality of the spin functions. In practice, weak bands in the spectra arising from such electronic transitions are frequently seen. However, they are at least an order of magnitude less intense than allowed transitions. One mechanism by which these transitions gain intensity is by spin-orbit coupling.

Intensity of the absorption bands for noncentrosymmetric complexes, *e.g.* tetrahedral, are much larger than those of centrosymmetric complexes.

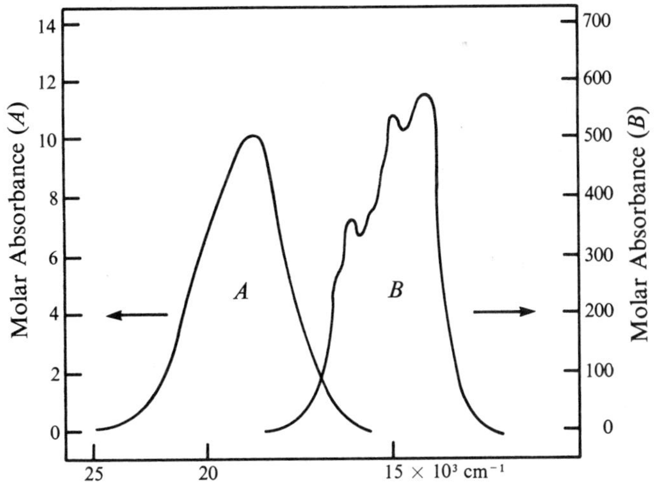

Figure 6.19
Comparison of the Spectra of $Co(H_2O)_6^{2+}$ *(A) and* $CoCl_4^{2-}$ *(B)*

A comparison of the spectra of $Co(H_2O)_6^{2+}$ and $CoCl_4^{2-}$, shown in Figure 6.19, emphasizes this point, since it can be seen that the intensity of the band for the tetrahedral complex is approximately 60 times that of the octahedral complex.

Colors of Compounds

The colors of compounds are determined by the wavelengths of light in the visible region that are absorbed. If there are no absorptions in the visible region, then the material is white or colorless depending upon the physical state. If all wave lengths are absorbed then the material is black. The relationship between the wave length absorbed and the color of the compound as detected by the eye, the complementary color, is summarized in Table 6.4. For example, the tris(ethylenediamine)cobalt(III) cation has a band in the visible region centered at 467 nm. From Table 6.4 we see that the complementary color is yellow. However, as shown in Figure 6.20, the band is rather broad, a considerable quantity of green-blue light is absorbed, and the color of $[Co(en)_3]^{3+}$ is orange.

Table 6.4
Absorbed and Complementary Colors

Wavelength region absorbed, mμ	*Color absorbed*	*Complementary color*
400–435	violet	yellow-green
435–480	blue	yellow
480–490	green-blue	orange
490–500	blue-green	red
500–560	green	purple
560–580	yellow-green	violet
580–595	yellow	blue
595–650	orange	green-blue
650–750	red	blue-green

Tanabe-Sugano Diagrams

Especially useful diagrams have been constructed by Sugano and Tanabe* for the $d^2 - d^8$ configurations and octahedral stereochemistry. In these diagrams, presented in Appendix 6D, the quantity B is a Racah parameter

*Y. Tanabe and S. Sugano, *J. Phys. Soc. Japan*, *9*, 753, 766 (1954).

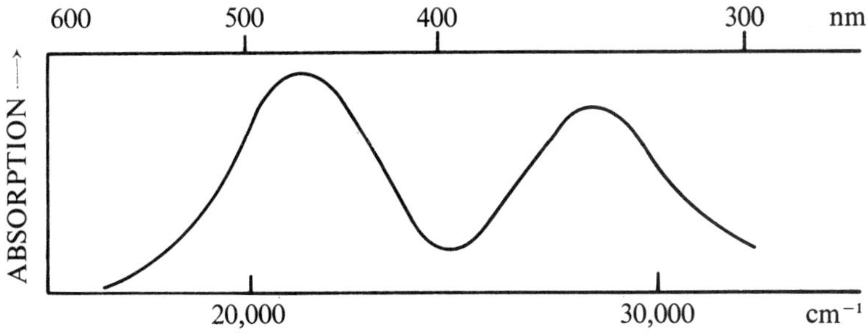

Figure 6.20
Absorption Spectrum of $Co(en)_3^{3+}$

which relates to interelectronic repulsion energies. These diagrams are very useful as guides for qualitative assignments of spectral transitions. Consider the spectrum of MnF_2 shown in Figure 6.21. The Mn^{2+} ion occupies a site in the crystal which has very nearly O_h symmetry. Therefore the diagram for d^5 may be used resulting in the assignments indicated in Figure 6.21. In fact, the diagrams may be used for molecules of considerably lower symmetry since absorption spectra are frequently recorded under such low resolution that some splittings can not be resolved.

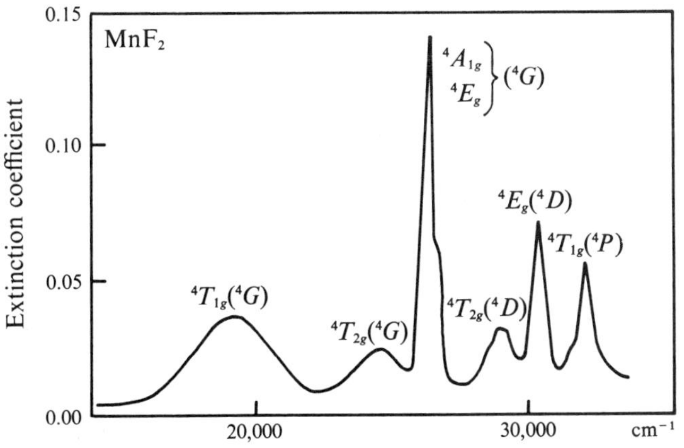

Figure 6.21
Absorption Spectrum of MnF_2

Also, as we pointed out in Section 6-6, the octahedral chart for d^n applies equally well to the tetrahedral d^{10-n} case.

Spectrochemical Series

Consider an ion with a d^1 electronic configuration. We have seen earlier that in an octahedral field the ^2D free ion term which arises from this configuration is split into $^2T_{2g}$ and ^2E_g, and the splitting increases as the strength of the crystal field increases. The energy of the absorption band corresponding to the excitation of an electron from the $^2T_{2g}$ to the ^2E_g state depends on the strength of the crystal field set up by the ligands. Extensive measurements on many compounds have led to an ordering of ligands in a spectrochemical series. The series is

$$\begin{aligned} &I^- < Br^- < SCN^- < Cl^- < NO_3^- < F^- < \text{urea} \simeq OH^- \simeq ONO^- \simeq HCOO^- \\ &< (C_2O_4)^{2-} < H_2O < NCS^- < \text{py} \simeq NH_3 < \text{en} < \text{dipy} < NO_2^- < CN^- \end{aligned} \tag{6-34}$$

The iodide ion as a ligand in this series gives rise to the weakest crystal field splitting (lowest value of 10 Dq) and the cyanide ion gives rise to the greatest splitting (highest value of 10 Dq). The series as given does not hold for every metal ion and oxidation state, but, by and large, gives a reasonable representation of experimental results.

There is also a spectrochemical series for the metal ions. In order of increasing Dq it is

$$\begin{aligned} Mn^{2+} < Ni^{2+} < Co^{2+} < Fe^{2+} < V^{2+} < Fe^{3+} < Cr^{3+} < V^{3+} \\ < Co^{3+} < Mo^{3+} < Rh^{3+} < Ru^{3+} < Pd^{4+} < Ir^{3+} < Pt^{4+} \end{aligned} \tag{6-35}$$

As expected from simple electrostatic considerations ions with higher oxidation numbers give rise to larger Dq values, as do elements in the second and third transition series. The latter observation is easy to understand since 4d and 5d orbitals have a greater radial expansion and these orbitals are expected to interact more strongly with ligands.

One final series merits mention. We have noted that covalency effects decrease the $^3F - {}^3P$ term splitting. The ratio

$$\beta = \frac{(^3F - {}^3P)\ \text{complex}}{(^3F - {}^3P)\ \text{free ion}}$$

has been defined as the nephelauxetic ratio, and is said to be indicative of covalency. This has been interpreted by Jorgensen to mean that the electrons occupy a larger volume of space in complexes with a consequent decrease in electron repulsions. Analogous ratios may be defined for metal ions. The series for some common ligands is

$$F^- < H_2O < \text{urea} < NH_3 < \text{en} < ox^{2-} < NCS^- < Cl^- < CN^- < N_3^- < Br^- < I^- \quad \textbf{(6-36)}$$

Some Selected Examples

It is now time to apply some of the concepts that have been developed to specific problems in transition metal spectroscopy. We will present and discuss a limited number of selected examples which demonstrate the utility of crystal field and ligand field theories and, perhaps, will show why these theories have captured the imagination of many chemists and physicists.

$Ti(H_2O)_6^{3+}$

The spectrum of the titanium(III) ion in dilute hydrochloric acid solution is shown in Figure 6.22. It is generally agreed that the chromophore responsible for the spectrum is $Ti(H_2O)_6^{3+}$. From our considerations in

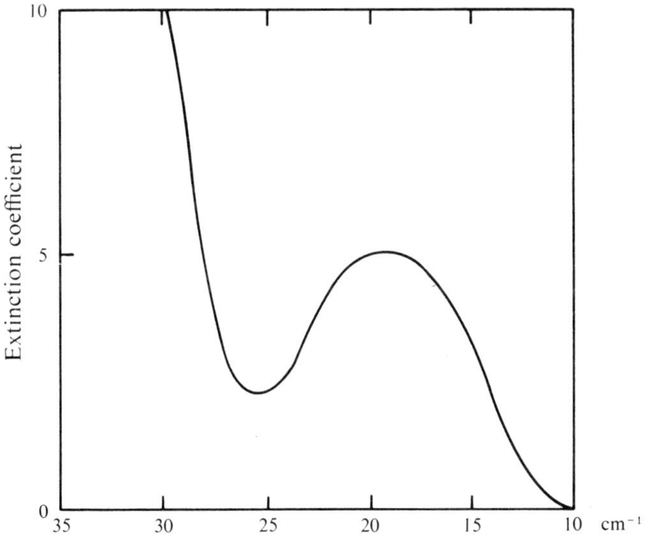

Figure 6.22
The Spectrum of $Ti(H_2O)_6^{3+}$

Section 6-4, we expect one band in the electronic spectrum to arise from the $^2T_{2g} \rightarrow {}^2E_g$ transition. However, it is apparent that there is not just one band in the spectrum but that there is a distinct splitting giving two bands. Analysis of the spectral bands reveals that the splitting is about 2,000 cm^{-1}. This splitting is the consequence of a structural distortion in agreement with the Jahn-Teller theorem, which states that structures giving rise to orbitally degenerate states will be unstable with respect to lower symmetry structures with orbitally nondegenerate states. For purposes of simplicity and in the absence of additional information one usually assumes that the distortional mode adopted is the one of highest symmetry which will remove the degeneracy. We have discussed such a distortional mode in Section 6-4, and the resultant energy level diagram is shown in Figure 6.6. A trigonal distortion will also remove the degeneracy of the ground state but will not remove the degeneracy of the E_g state.

We call attention to the intensity of the spectral band in $Ti(H_2O)_6^{3+}$, which is the order of magnitude of *d-d* bands in centrosymmetric complexes of first row transition metal ions.

$Cr(ox)_3^{3-}$

The use of polarized light is very important for the assignment of spectral transitions. To use the technique it is usually necessary to obtain suitable single crystal samples, although recently there have been some spectra of oriented molecules recorded using liquid crystals. The polarized spectrum shown in Figure 6.23 of a single crystal of $NaMgAl(C_2O_4)_3 \cdot 9H_2O$ into which chromium ions were isomorphously substituted for aluminum illustrates the use of selection rules for the assignment of spectral bands*. The crystal is uniaxial, and the trigonal axis of the $M(ox)_3^{3-}$ ions (see Figure 6.24) are aligned parallel with the *c* axis of the crystal.

In the experiment the single crystal sample is oriented.in the spectrometer such that the trigonal axis is perpendicular to the propagation vector of the radiation beam, and a suitable polarizer, say a sheet of Polaroid film, is also mounted perpendicular to the propagation vector. By careful rotation of the polarizer it is possible to orient the plane of polarized light passed by the polarizer such that the plane is parallel with or perpendicular to the trigonal axis of the crystal, and, therefore parallel with or perpendicular to the trigonal axis of the $M(ox)_3^{3-}$ ions. When the plane of polarized light is parallel with the trigonal axis, only the z component of the electric dipole operator can enable the transition, and the resultant spectrum is called the parallel ($||$) spectrum. When the polarization is perpendicular to the trigonal axis, the (x, y) components only enable the

*T. S. Piper and R. L. Carlin, *J. Chem. Phys.*, *35*, 1809 (1961).

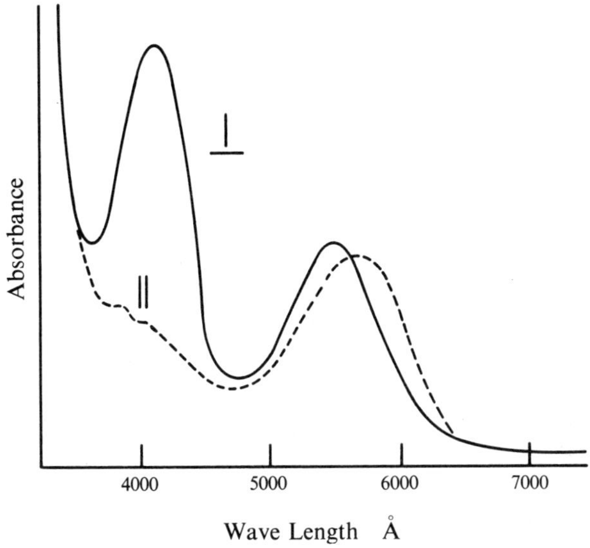

Figure 6.23
Polarized Crystal Spectrum of $Cr(ox)_3^{3-}$ *in* Na Mg $Al(C_2O_4)_3 \cdot 9H_2O$

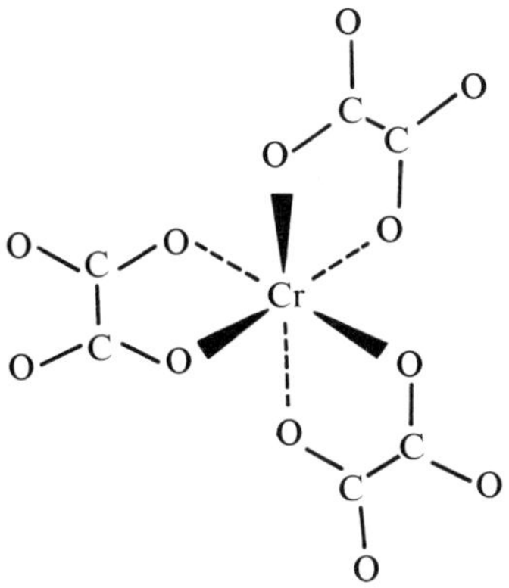

Figure 6.24
Structure of $Cr(ox)_3^{3-}$. *The Trigonal Axis is Perpendicular to the Plane of the Page*

transition, and the spectrum is called the perpendicular spectrum ($\perp$). Examination of Figure 6.23 will show that the parallel ($||$) and perpendicular ($\perp$) spectra of $Cr(ox)_3^{3-}$ are remarkably different. We will now show how the spectra may be understood in terms of selection rules. By reference to the Tanabe-Sugano diagram for the d^3 configuration we find the following array of quartet spin states:

$$^4A_{2g},\ ^4T_{2g},\ ^4T_{1g}(^4F),\ ^4T_{1g}(^4P)$$

Since only two major bands are shown in Figure 6.23, apparently the transition to the $^4T_{1g}(^4P)$ state was not detected.

The local symmetry of the chromium ion is D_3, so the following correlations exist:

O_h	D_3
$^4A_{2g}$	$^4A_2$
$^4T_{2g}$	$^4A_1 + {}^4E$
$^4T_{1g}$	$^4A_2 + {}^4E$

That is, in the lower symmetry complex some of the degeneracies of the triply degenerate states are removed giving a singly degenerate and a doubly degenerate level in each case. However, some of the transitions may not be allowed, and in addition, they may be polarized, i.e., be allowed only in the $\perp$ or the $||$ polarization, but not necessarily in both polarizations. The possible dipole integrals and the direct products of Ψ_e^{excited} $\mu(x, y, z)\,\Psi_e^{\text{ground}}$ are

$$\int \Psi(^4A_1)\,\mu_z\,\Psi(^4A_2)\,d\tau \qquad A_1 \times A_2 \times A_2 = A_1$$

$$\int \Psi(^4A_1)\,\mu_{x,y}\,\Psi(^4A_2)\,d\tau \qquad A_1 \times E \times A_2 = E$$

$$\int \Psi(^4E)\,\mu_z\,\Psi(^4A_2)\,d\tau \qquad E \times A_2 \times A_2 = E$$

$$\int \Psi(^4E)\,\mu_{x,y}\,\Psi(^4A_2)\,d\tau \qquad E \times E \times A_2 = A_1 + A_2 + E$$

$$\int \Psi(^4A_2)\,\mu_z\,\Psi(^4A_2)\,d\tau \qquad A_2 \times A_2 \times A_2 = A_2$$

$$\int \Psi(^4A_2)\,\mu_{x,y}\,\Psi(^4A_2)\,d\tau \qquad A_2 \times E \times A_1 = E$$

Therefore, when the radiation beam is polarized parallel to the trigonal axis the following transition is allowed:

$$^4A_2 \longrightarrow {}^4A_1$$

In the perpendicular polarization the following transitions are allowed:

$$^4A_2 \longrightarrow {}^4E\,({}^4T_{2g})$$
$$^4A_2 \longrightarrow {}^4E\,({}^4T_{1g})$$

The assignment of the bands in polarized spectra is now possible. The || spectrum shows one intense band at ~5800 Å which must be the $^4A_2 \longrightarrow {}^4A_1\,({}^4T_{2g})$ transition, while the two bands in the $\perp$ spectrum are the transitions to the ^4E components of $^4T_{2g}$ and $^4T_{1g}$. The $^4A_{2g} \longrightarrow {}^4A_{2g}({}^4T_{1g})$ is not allowed by the selection rules, and the absorption in the || spectrum at the expected energy is very weak.

Exercise 6-4

One might argue that the chelate cycles have no effect on the spectral properties of the complex and that an analysis of the spectrum in terms of a CrO_6 chromophore with symmetry D_{3d} would be more appropriate. Since there is a center of inversion in D_{3d}, the transition would be governed by vibronic selection rules. Work out the vibronic rules selection and comment on the alternate argument presented here.

$CuCl_4^{2-}$

Tetrahedral d^9 complexes have the configuration $(e)^4(t_2)^5$ and in the absence of other effects that may remove the degeneracy are subject to the Jahn-Teller effect. All salts of the $CuCl_4^{2-}$ ion that have been studied by x-ray diffraction are distorted, and a typical energy level diagram for $CuCl_4^{2-}$ has been given in Figure 6-16. In agreement with this expected energy level scheme, as shown in Figure 6.25, four bands are observed in the spectrum of the $CuCl_4^{2-}$ ion contained as a guest in the host lattice Cs_2ZnCl_4. The copper ion occupies a site with C_s symmetry, but to a good approximation the symmetry of $CuCl_4^{2-}$ is D_{2d}. The energies and assignments in D_{2d} are

$^2B_2 \longrightarrow {}^2E$	5550, 4800 cm^{-1}
$^2B_2 \longrightarrow {}^2B_1$	7900
$^2B_2 \longrightarrow {}^2A_1$	9050

Note that two transitions to the ^2E level were seen. The two states arise from a splitting of the ^2E by spin-orbit coupling and the low symmetry component of the crystal field.

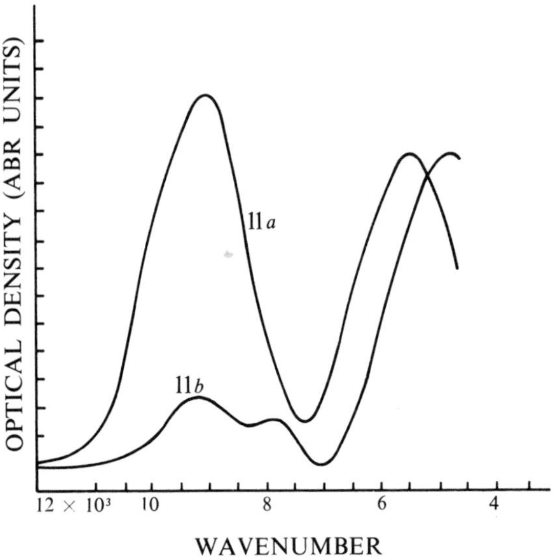

Figure 6.25
Polarized Spectra of Cs_2CuCl_4 *at 20°K. The Incident Radiation Beam is Perpendicular to the (001) Face, and the Two Polarizations Shown are Parallel to the a and b Axes of the Orthorhombic Crystal. The Mirror Plane of Symmetry of the* $CuCl_4^{2-}$ *Ion is Perpendicular to the b Axis.* (Adapted from J. Ferguson, *J. Chem Phys.*, 40, 3406, (1964)).

6-10 MAGNETISM

By definition the magnetic moment of an energy level, E_n, is

$$\mu_n = -\frac{\partial E_n}{\partial H}$$

where H is the magnetic field, and the total moment, M, of the system is given by

$$M = \frac{N \sum_n \mu_n \exp(-E_n/kT)}{\sum_n \exp(-E_n/kT)}$$

where N is Avogadro's number, and the summation is over all the states. The energy of the nth level can be expanded in a power series

$$E_n = E_n^0 + HE_n^{(1)} + H^2E_n^{(2)} + \ldots$$

where E_n^0 is the zero field energy, and $E_n^{(1)}$ and $E_n^{(2)}$ are the first and second order Zeeman coefficients. Thus,

$$\mu_n = -E_n^{(1)} - 2HE_n^{(2)} + \ldots$$

If we note that

$$\exp(-E_n/kT) = \exp\left[\frac{-E_n^0 - HE_n^{(1)} - H^2E_n^{(2)} - \ldots}{kT}\right]$$

$$\simeq \left[1 - \frac{HE_n^{(1)}}{kT}\right]\exp\left(\frac{-E_n^0}{kT}\right)$$

then

$$M = \frac{N\sum_n(-E_n^{(1)} - 2HE_n^{(2)} - \ldots)(1 - HE_n^{(1)}/kT)\exp(-E_n^0/kT)}{\sum_n(1 - HE_n^{(1)}/kT)\exp(-E_n^0/kT)} \qquad \textbf{(6-37)}$$

Paramagnetic substances do not exhibit moments in the absence of magnetic fields so

$$\sum_n\{-E_n^{(1)}\exp(-E_n^0/kT)\} = 0$$

At moderate temperatures and magnetic fields only first order terms in H in the numerator and only those independent of H in the denominator need be retained, and Equation (6-38) results. This equation was derived by Van Vleck and bears his name. Note that $\chi = M/H$.

$$\chi_m = \frac{N\sum_n[(E_n^{(1)})^2/kT - 2E_n^{(2)}]\exp(-E_n^0/kT)}{\sum_n\exp(-E_n^0/kT)} \qquad \textbf{(6-38)}$$

Thus, the calculation of magnetic properties involves the determination of the first and second order Zeeman coefficients and substitution into the Van Vleck equation.

The operator for the Zeeman effect is

$$\mu = \beta\vec{H}\cdot(\vec{L_i} + 2\vec{S_i}) \qquad \textbf{(6-39)}$$

where i designates one of the three directions x, y, and z, and the first and second order Zeeman energies become

$$E_n^{(1)} = \langle \Psi_n | \mu | \Psi_n \rangle$$
$$E_n^{(2)} = \sum_{j,i \neq j} |\langle \Psi_i | \mu | \Psi_j \rangle|^2/(E_i - E_j)$$

We shall first calculate the first order Zeeman energies for the $\mathbf{\Gamma}_7$ and $\mathbf{\Gamma}_8$ spin-orbit levels (Equation (6-31)), which were seen to arise from the ${}^2T_{2g}$ state. The magnetic moment operator reduces to

$$(L_z + 2S_z)\beta H_z$$

and the calculation of

$$\langle \Psi_1 | L_z + 2S_z | \Psi_1 \rangle \beta H_z$$

is shown in detail:

$$\left\langle \frac{1}{\sqrt{3}} \left[\frac{1}{\sqrt{2}} (d_2 - d_{-2})\beta - \sqrt{2}\, d_1 \alpha \right] \right| (L_z + 2S_z) \left| \frac{1}{\sqrt{3}} \left[\frac{1}{\sqrt{2}} (d_2 - d_{-2})\beta - \sqrt{2}\, d_1 \alpha \right] \right\rangle \beta H_z$$
$$= \tfrac{1}{3}\{\tfrac{1}{2}\langle (d_2 - d_{-2}) | L_z | (d_2 - d_{-2}) \rangle \langle \beta | \beta \rangle + 2 \langle d_1 | L_z | d_1 \rangle \langle \alpha | \alpha \rangle$$
$$+ \tfrac{1}{2}\langle (d_2 - d_{-2}) | (d_2 - d_{-2}) \rangle \langle \beta | 2S_z | \beta \rangle + 2\langle d_1 | d_1 \rangle \langle \alpha | 2S_z | \alpha \rangle\} \beta H_z$$
$$= \beta H_z$$

The other results are

$$\langle \Psi_2 | \mu | \Psi_2 \rangle = -\beta H_z$$
$$\langle \Psi_3 | \mu | \Psi_3 \rangle = 0$$
$$\langle \Psi_4 | \mu | \Psi_4 \rangle = 0$$
$$\langle \Psi_5 | \mu | \Psi_5 \rangle = 0$$
$$\langle \Psi_6 | \mu | \Psi_6 \rangle = 0$$

We have now shown that the $\mathbf{\Gamma}_7$ level is split by the external magnetic field but that $\mathbf{\Gamma}_8$ is nonmagnetic.

For the second order Zeeman effect, calculations show that

$$\langle \Psi_1 | \mu | \Psi_3 \rangle = \langle \Psi_3 | \mu | \Psi_1 \rangle = \sqrt{2}\, \beta H$$
$$\langle \Psi_2 | \mu | \Psi_4 \rangle = \langle \Psi_4 | \mu | \Psi_2 \rangle = \sqrt{2}\, \beta H$$

and

$$E_1^{(2)} = (\sqrt{2}\,\beta H)^2/[\lambda - (-\lambda/2)] = 4\beta^2 H^2/3\lambda$$
$$E_2^{(2)} = 4\beta^2 H^2/3\lambda$$
$$E_3^{(2)} = (\sqrt{2}\,\beta H)^2/[-\lambda/2 - \lambda] = -4\beta^2 H^2/3\lambda$$
$$E_4^{(2)} = -4\beta^2 H^2/3\lambda$$
$$E_5^{(2)} = 0$$
$$E_6^{(2)} = 0$$

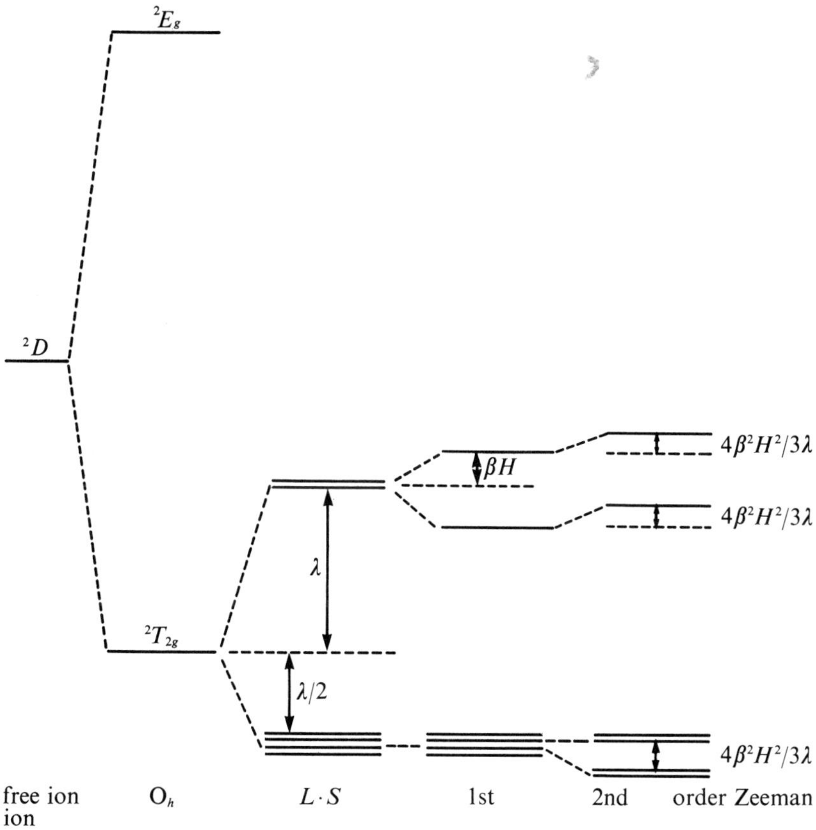

Figure 6.26

Energy Level Diagram for ^2D Term in O_h Field Showing Influence of Spin-Orbit Coupling and an External Magnetic Field

The final energy level diagram which summarizes these calculations is given in Figure 6.26. Substitution into the Van Vleck equation yields:

$$M = \frac{N\{2(0 + 8\beta^2/3\lambda) \exp(\lambda/2kT) + 2(\beta^2/kT - 8\beta^2/3\lambda) \exp(-\lambda/kT)}{4 \exp(\lambda/2kT) + 2 \exp(-\lambda/kT)}$$

If we multiply the numerator and denominator by $\exp(-\lambda/2kT)$ and divide out a factor of 2 there results

$$\chi_M = \frac{N\{8\beta^2/3\lambda + (\beta^2/kT - 8\beta^2/3\lambda) \exp(-3\lambda/2kT)\}}{2 + \exp(-3\lambda/2kT)}$$

Taking note of the definition

$$\mu_{\text{eff}}^2 = \frac{3kT}{N\beta^2} \chi_M$$

then

$$\mu_{\text{eff}}^2 = \frac{8 + [3(\lambda/kT) - 8] \exp(-3\lambda/2kT)}{(\lambda/kT)\{2 + \exp(-3\lambda/2kT)\}}$$

The temperature variation of μ_{eff} is shown in Figure 6.27. For this system the moment approaches zero as the temperature approaches zero even though there is one unpaired electron.

There is much to be learned about the electronic structures of paramagnetic compounds from a study of the temperature variation of the magnetic properties. However, we will terminate our discussions at this

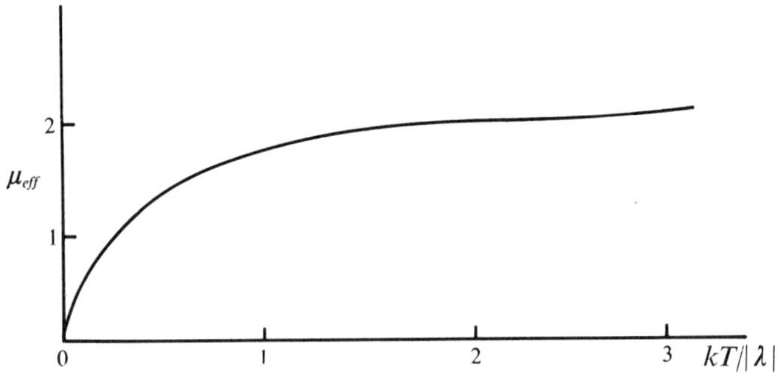

Figure 6.27
Variation of μ_{eff} with Temperature for the $^2T_{2g}$ State

point and refer the reader to the more complete treatments in the standard reference works.

Bibliography

C. J. Ballhausen, *Introduction to Ligand Field Theory*, New York: McGraw-Hill Book Co., 1962.

T. M. Dunn, D. S. McClure, and R. G. Pearson, *Some Aspects of Crystal Field Theory*, New York: Harper and Row Publishers, 1965.

H. Eyring, J. Walter, and G. E. Kimbal, *Quantum Chemistry*, New York: John Wiley and Sons, Inc., 1944.

B. N. Figgis, *Introduction to Ligand Fields*, New York: Interscience Publishers, 1966.

J. S. Griffith, *The Theory of Transition Metal Ions*, Cambridge: University Press, 1961.

C. K. Jorgensen, *Absorption Spectra and Chemical Bonding in Complexes*, New York: Pergamon Press, 1962.

Appendix 6A

Microstate Chart for d^2 Configuration

	M_S		
M_L	1	0	−1
4		$(\overset{+}{2}, \overset{-}{2})$	
3	$(\overset{+}{2}, \overset{+}{1})$	$(\overset{+}{2}, \overset{-}{1})(\overset{-}{2}, \overset{+}{1})$	$(\overset{-}{2}, \overset{-}{1})$
2	$(\overset{+}{2}, \overset{+}{0})$	$(\overset{+}{2}, \overset{-}{0})(\overset{-}{2}, \overset{+}{0})(\overset{+}{1}, \overset{-}{1})$	$(\overset{-}{2}, \overset{-}{0})$
1	$(\overset{+}{1}, \overset{+}{0})(\overset{+}{2}, -\overset{+}{1})$	$(\overset{+}{1}, \overset{-}{0})(\overset{-}{1}, \overset{+}{0})$ $(\overset{+}{2}, -\overset{-}{1})(\overset{-}{2}, -\overset{+}{1})$	$(\overset{-}{1}, \overset{-}{0})(\overset{-}{2}, -\overset{-}{1})$
0	$(\overset{+}{2}, -\overset{+}{2})(\overset{+}{1}, -\overset{+}{1})$	$(\overset{+}{2}, -\overset{-}{2})(\overset{-}{2}, -\overset{+}{2})$ $(\overset{+}{1}, -\overset{-}{1})$ $(\overset{-}{1}, -\overset{+}{1})(\overset{+}{0}, \overset{-}{0})$	$(\overset{-}{2}, -\overset{-}{2})(\overset{-}{1}, -\overset{-}{1})$
−1	$(-\overset{+}{1}, \overset{+}{0})(\overset{+}{1}, -\overset{+}{2})$	$(-\overset{+}{1}, \overset{-}{0})(-\overset{-}{1}, \overset{+}{0})$ $(-\overset{+}{2}, \overset{-}{1})(-\overset{-}{2}, \overset{+}{1})$	$(-\overset{-}{1}, \overset{-}{0})(\overset{-}{1}, -\overset{-}{2})$
−2	$(-\overset{+}{2}, \overset{+}{0})$	$(-\overset{+}{2}, \overset{-}{0})(-\overset{-}{2}, \overset{+}{0})(-\overset{+}{1}, -\overset{-}{1})$	$(-\overset{-}{2}, \overset{-}{0})$
−3	$(-\overset{+}{2}, -\overset{+}{1})$	$(-\overset{+}{2}, -\overset{-}{1})(-\overset{-}{2}, -\overset{+}{1})$	$(-\overset{-}{2}, -\overset{-}{1})$
−4		$(-\overset{+}{2}, -\overset{-}{2})$	

The following Russell-Saunders terms may be obtained from the chart: ${}^3F$, ${}^3P$, ${}^1G$, ${}^1D$, ${}^1S$.

Appendix 6B

Correlation Diagram for d^2 Ion in C_{4v} Environment

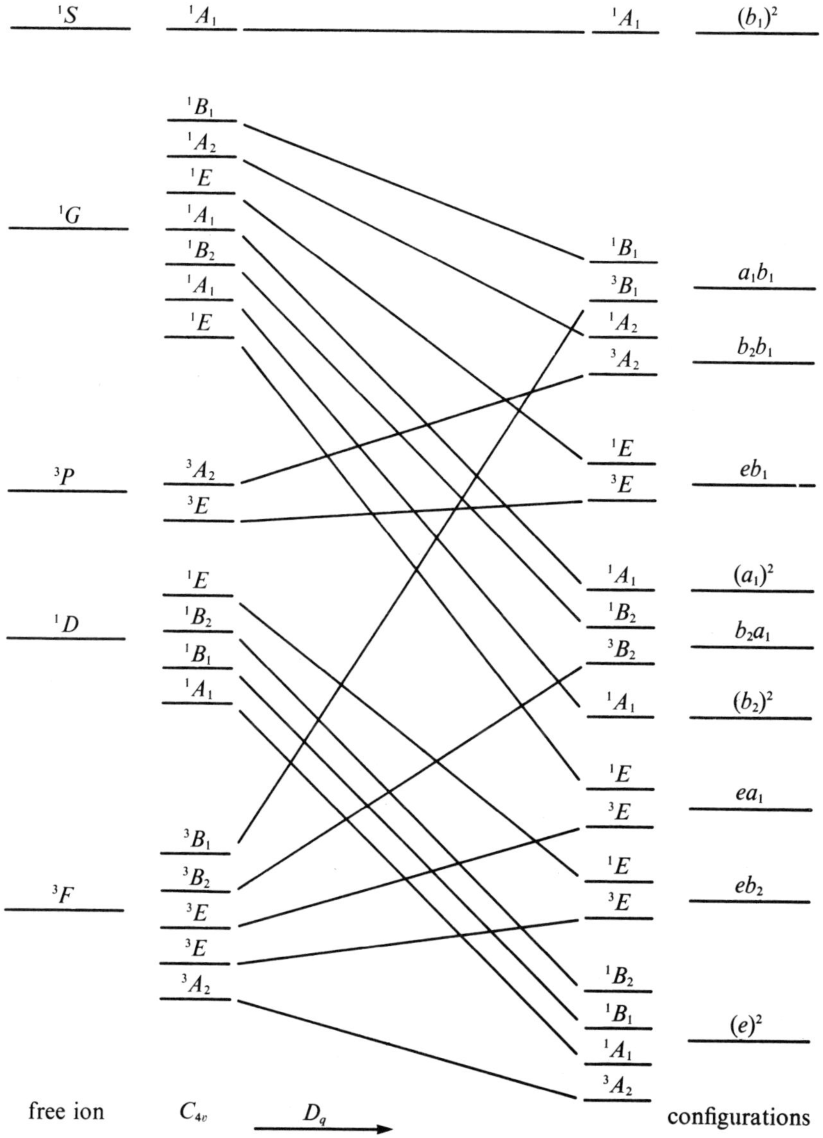

Appendix 6C

Character Tables for Some Useful Double Groups

D_{2d}'	E	R	$S_4{}^3R$ S_4	S_4R S_4^3	C_2R C_2	$2C_2'R$ $2C_2'$	$2\sigma_a$
$\Gamma_1 A_1$	1	1	1	1	1	1	1
$\Gamma_2 A_2$	1	1	1	1	1	−1	−1
$\Gamma_3 B_1$	1	1	−1	−1	1	1	−1
$\Gamma_4 B_2$	1	1	−1	−1	1	−1	1
$\Gamma_5 E_1$	2	2	0	0	−2	0	0
$\Gamma_6 E_2$	2	−2	$\sqrt{2}$	$-\sqrt{2}$	0	0	0
$\Gamma_7 E_3$	2	−2	$-\sqrt{2}$	$\sqrt{2}$	0	0	0

D_4'	E	R	C_4 C_4^3R	C_4^3 C_4R	C_2 C_2R	$2C_2'$ $2C_2'R$	$2C_2''$ $2C_2''R$
$\Gamma_1 A_1'$	1	1	1	1	1	1	1
$\Gamma_2 A_2'$	1	1	1	1	1	−1	−1
$\Gamma_3 B_1'$	1	1	−1	−1	1	1	−1
$\Gamma_4 B_2'$	1	1	−1	−1	1	−1	1
$\Gamma_5 E_1'$	2	2	0	0	−2	0	0
$\Gamma_6 E_2'$	2	−2	$\sqrt{2}$	$-\sqrt{2}$	0	0	0
$\Gamma_7 E_3'$	2	−2	$-\sqrt{2}$	$\sqrt{2}$	0	0	0

O'	E	R	$4C_3$ $4C_3^2R$	$4C_3^2$ $4C_3R$	$3C_2$ $3C_2R$	$3C_4$ $3C_4^3R$	$3C_4^3$ $3C_4R$	$6C_2'$ $6C_2'R$
$\Gamma_1 A_1'$	1	1	1	1	1	1	1	1
$\Gamma_2 A_2'$	1	1	1	1	1	−1	−1	−1
$\Gamma_3 E_1'$	2	2	−1	−1	2	0	0	0
$\Gamma_4 T_1'$	3	3	0	0	−1	1	1	−1
$\Gamma_5 T_2'$	3	3	0	0	−1	−1	−1	1
$\Gamma_6 E_2'$	2	−2	1	−1	0	$\sqrt{2}$	$-\sqrt{2}$	0
$\Gamma_7 E_3'$	2	−2	1	−1	0	$-\sqrt{2}$	$\sqrt{2}$	0
$\Gamma_8 G'$	4	−4	−1	1	0	0	0	0

Reprinted with permission from F. A. Cotton, *Chemical Applications of Group Theory*, New York: Wiley-Interscience, 2 nd ed., 1971.

Appendix 6D

Tanabe-Sugano Diagrams for Metal Ions with Configurations, d^2-d^6

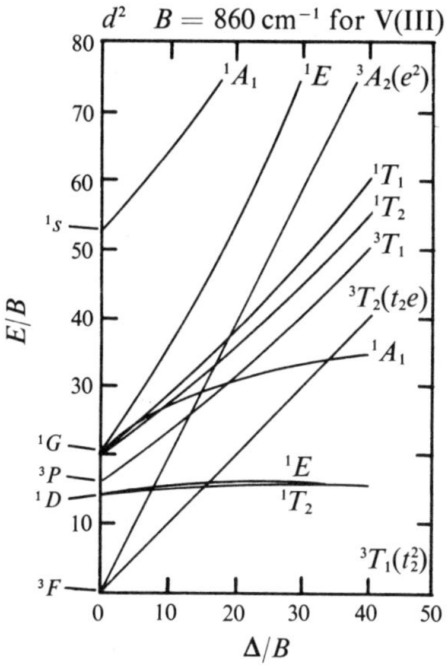

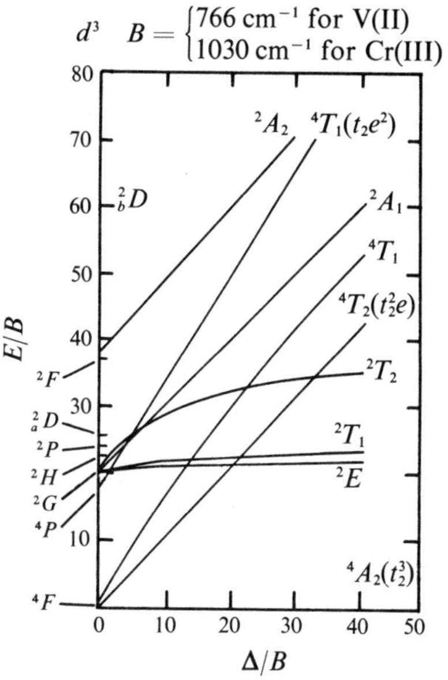

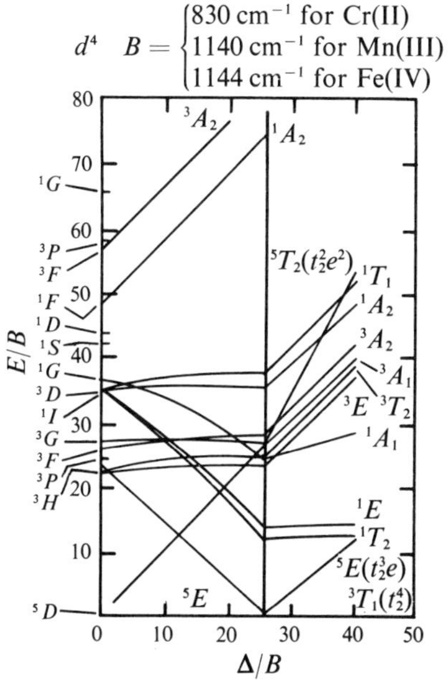

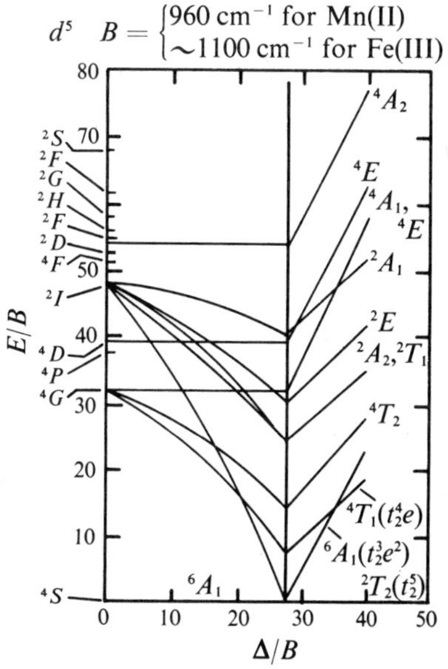

$d^6 \quad B = \begin{cases} 1058\ \text{cm}^{-1} \text{ for Fe(II)} \\ 1065\ \text{cm}^{-1} \text{ for Co(III)} \end{cases}$

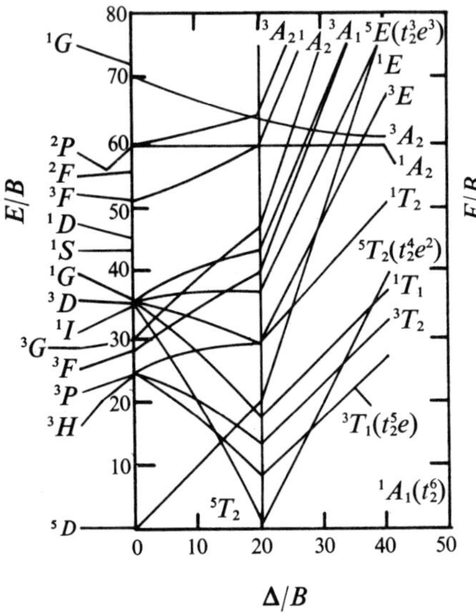

$d^7 \quad B = 970\ \text{cm}^{-1}$ for Co(II)

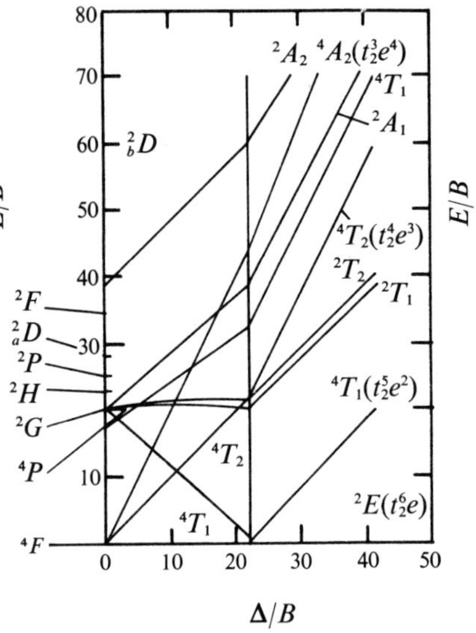

$d^8 \quad B = 1080\ \text{cm}^{-1}$ for Ni(II)

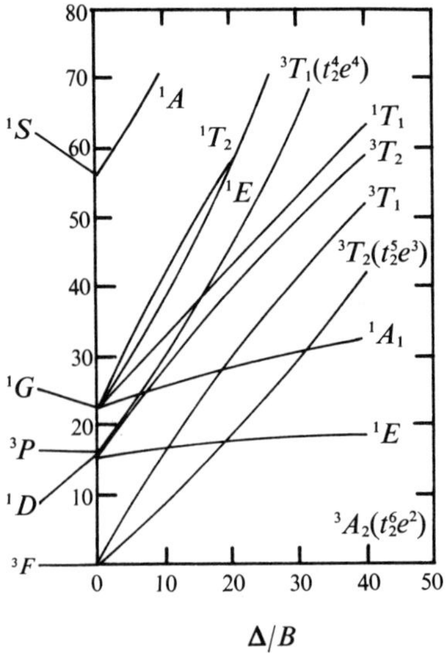

7

MOLECULAR ORBITAL THEORY

7-1 THE HYDROGEN MOLECULE ION

In molecular orbital theory electrons are considered to be in orbitals which are associated with several nuclei. These molecular orbitals are not simple atomic orbitals, but are usually approximated as some combination of atomic orbitals. A common approximation is a simple linear combination of atomic orbitals (LCAO). These LCAO's are conveniently generated for complex molecules by group theoretical techniques.

An example dealing with the simplest possible molecule, the hydrogen molecule-ion, should illustrate these ideas. As shown in Figure 7.1, if the two nuclei of the H_2^+ ion are placed at the observed internuclear separation, the hydrogen $1s$ wave functions overlap each other. Thus, a reasonable approximation to a molecular orbital for H_2^+ would be that the electron is in the $1s$ orbital of H_A when it is near H_A and in the $1s$ orbital of H_B when it is near H_B. The approximation is questionable in the region of orbital overlap.

Let us consider the molecular orbital as a linear combination of the two $1s$ atomic orbitals. We may write this molecular orbital as $\phi = a_1\psi_a + a_2\psi_b$ where $a_1 = a_2$ since both atomic orbitals are identical. Also, the molecular orbital must be normalized, where the normalization condition

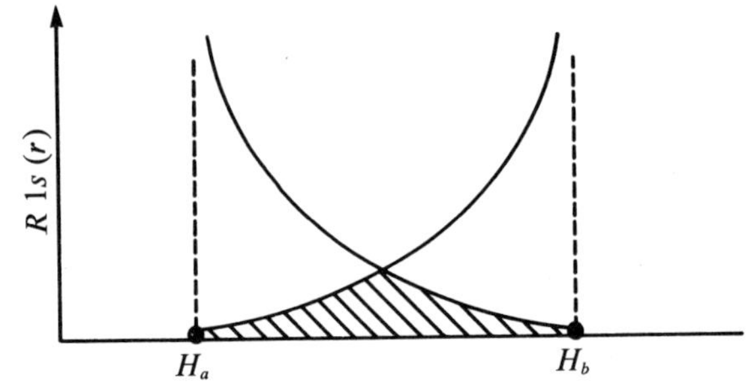

Figure 7.1
Overlap in the H_2^+ *Ion*

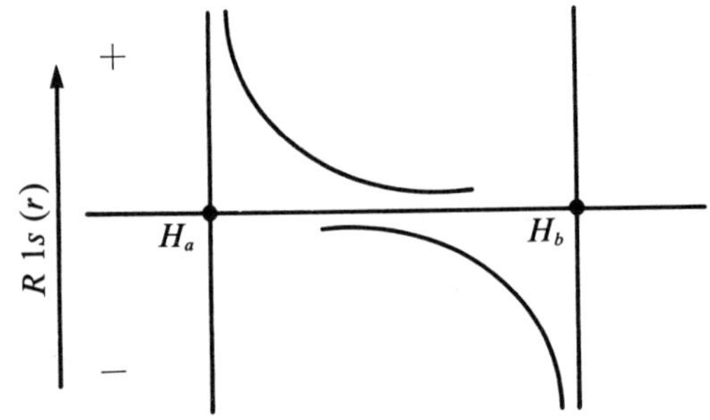

Figure 7.2
Minimized Overlap in H_2^+

is $N^2 \int \phi^* \phi \, d\tau = 1$. We may also take the other possible linear combination of the two atomic orbitals, i.e., $\phi = a_1\psi_a - a_2\psi_b$, because we have combined two atomic orbitals, and to conserve the number of orbitals, two molecular orbitals must be constructed. The picture of this negative linear combination is shown in Figure 7.2, where it can be seen that density is minimized between the two hydrogen atoms. We will normalize the function $\phi = N(\psi_a + \psi_b)$ by solving for N. Thus,

$$N^2 \int (\psi_a + \psi_b)^2 \, d\tau = N^2 \left(\int \psi_a^2 \, d\tau + \int \psi_b^2 \, d\tau + 2 \int \psi_a \psi_b \, d\tau \right) = 1.$$

Since the atomic orbitals which were taken to make the LCAO are assumed to be normalized, then $\int \psi_a^2 \, d\tau = \int \psi_b^2 \, d\tau = 1$ and $\int \psi_a \psi_b \, d\tau$ is the overlap integral, S_{ab}. Thus,

$$N^2 = 1/(2 + 2\,S_{ab}) \text{ and } N = \pm\sqrt{\frac{1}{2 + 2\,S_{ab}}}$$

In many instances the atomic overlap is neglected because it is small. If this is assumed, then $N = \sqrt{1/2}$, and the two molecular orbitals are shown in Equation (7-1).

$$\psi_\sigma = (1/\sqrt{2})(\psi_a + \psi_b) \qquad \textbf{(7-1a)}$$

$$\psi_{\sigma^*} = (1/\sqrt{2})(\psi_a - \psi_b) \qquad \textbf{(7-1b)}$$

The upper limits of the true energies of the molecular orbitals can be found using the Schrödinger Equation $H\psi = E\psi$ and these approximate wave functions. If we multiply both sides of this equation by ψ^* and integrate, it becomes $\int \psi^* H\psi\, d\tau = E \int \psi^2\, d\tau = E$. Equation (7-2) shows an

$$E(\psi_\sigma) = \int \psi_\sigma H\psi_\sigma\, d\tau = \tfrac{1}{2}\int (\psi_a + \psi_b)H(\psi_a + \psi_b)\, d\tau$$

$$= \tfrac{1}{2}\int \psi_a H\psi_a\, d\tau + \tfrac{1}{2}\int \psi_b H\psi_a\, d\tau + \tfrac{1}{2}\int \psi_a H\psi_b\, d\tau + \tfrac{1}{2}\int \psi_b H\psi_b\, d\tau \qquad \textbf{(7-2)}$$

expression for the energy of the molecular orbital in Equation (7-1a), which can be simplified by the introduction of the following definitions:

$\int \psi_a \mathrm{H}\psi_a\, d\tau = \mathrm{H}_{AA}$, which is called the coulomb integral

$\int \psi_a H\psi_b\, d\tau = H_{AB}$, which is called the resonance or exchange integral.

Simplifying Equation (7-2), we obtain $E(\psi_\sigma) = \frac{1}{2}\,\mathrm{H}_{AA} + \frac{1}{2}\,\mathrm{H}_{BA} + \frac{1}{2}\,\mathrm{H}_{AB} + \frac{1}{2}\,\mathrm{H}_{BB}$, but for the H_2 molecule $\mathrm{H}_{AA} = \mathrm{H}_{BB}$ and $\mathrm{H}_{AB} = \mathrm{H}_{BA}$, so we have $E(\psi_\sigma) = \mathrm{H}_{AA} + \mathrm{H}_{AB}$. The energy of the other molecular orbital (Equation 7-1b) can be calculated in a similar manner:

$$E(\psi_\sigma^*) = \tfrac{1}{2}\int (\psi_a - \psi_b)^* H(\psi_a - \psi_b)\, d\tau = \mathrm{H}_{AA} - \mathrm{H}_{AB}.$$

These two energies have been calculated assuming that the overlap between hydrogen atoms (S) equals zero. It is left as an exercise for the reader to calculate the two energy levels in terms of H_{AA} and H_{AB} assuming that $S \neq 0$. Figure 7.3 shows the result of the above calculation. Each 1s orbital has an energy equal to the coulombic energy of that particular orbital. One molecular orbital (Equation (7-1a)) is more stable than the atomic orbitals and is called the bonding molecular orbital, while the molecular orbital (Equation (7-1b)) that is more unstable is called the antibonding

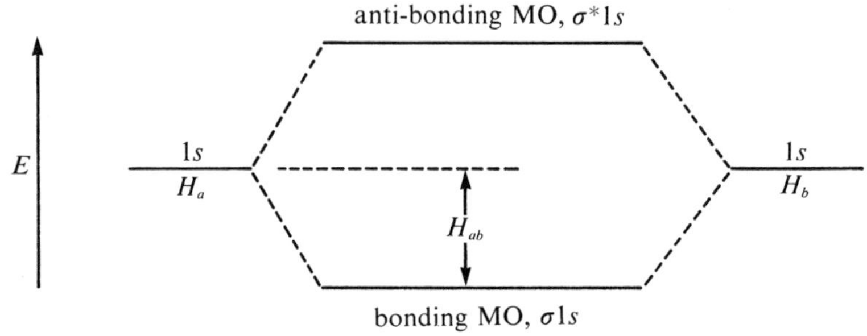

Figure 7.3
The Molecular Orbitals of H_2^+

molecular orbital. The energy difference between the bonding and antibonding molecular orbitals is equal to $2H_{AB}$. The one electron that belongs to the H_2^+ molecule-ion will occupy the bonding molecular orbital, which is lowest in energy, and this electron provides the density which binds the two nuclei together.

7-2 MOLECULAR AUFBAU PRINCIPLE

The aufbau or "building up" principle can be applied to the electron occupation of molecular orbitals just as it was applied to the occupation of atomic orbitals. The Pauli Exclusion Principle and Hund's Rules are both applicable to molecular orbitals. Thus, the two molecular orbitals constructed from the two 1*s* atomic orbitals can accommodate 4 electrons. For the H_2 molecule, both electrons occupy the $\sigma 1s$ bonding molecular orbital, giving the configuration $(\sigma 1s)^2$.

If two electrons occupy a bonding molecular orbital, a bond is formed, and if two electrons occupy an antibonding molecular orbital, a bond is subtracted. With this in mind we define the *bond order* to be (the number of electrons in bonding MO's—the number of electrons in antibonding MO's)/2. Therefore, in the H_2 molecule with 2 electrons in $\sigma 1s$, the bond order is one. For a molecule such as He_2 with 4 electrons, the bond order is zero because 2 electrons are in $\sigma 1s$ and 2 electrons are in $\sigma^* 1s$, i.e., He_2 would have the configuration $(\sigma 1s)^2(\sigma^* 1s)^2$. In order to deal with homonuclear diatomic molecules containing more than 4 electrons we must first construct more molecular orbitals.

The molecular orbitals constructed from two 2*s* atomic orbitals would be very similar to what we obtained using 1*s* atomic orbitals, the primary

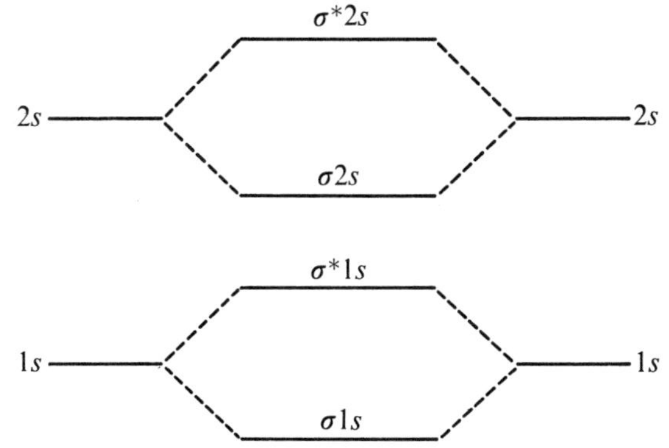

Figure 7.4
The Molecular Orbitals formed from 1s and 2s Atomic Orbitals

difference being that the 2*s* atomic orbitals are higher in energy. In fact, the $\sigma 2s$ molecular orbital would be higher in energy than the $\sigma^* 1s$ molecular orbital, as shown in Figure 7.4.

Molecular orbitals constructed from 2*p* atomic orbitals present a different problem because 2*p* orbitals are capable of σ and π bonding. In the diatomic molecule we shall designate the *z*-axis on each atom as the axis along which the σ bond lies. Figure 7.5 shows two of the *p* orbitals and their relative bonding capabilities. In the formation of molecular orbitals from the 2*p* atomic orbitals we will first deal with the σ bonding orbitals. Because there are two (p_z^1, p_z^2) atomic orbitals involved, two molecular orbitals must be constructed. In Figure 7.6 we can see the result of com-

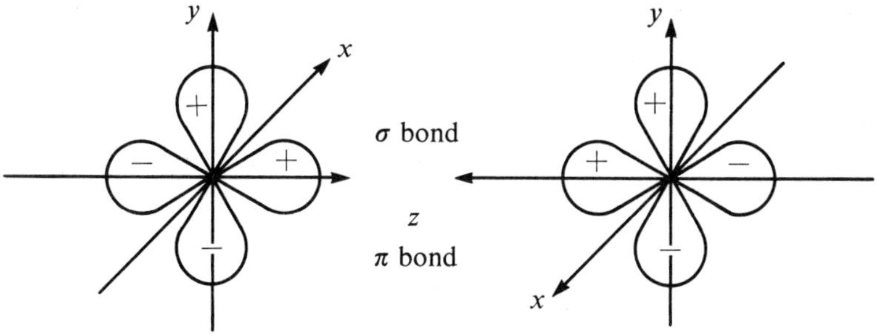

Figure 7.5
P Orbital Bonding

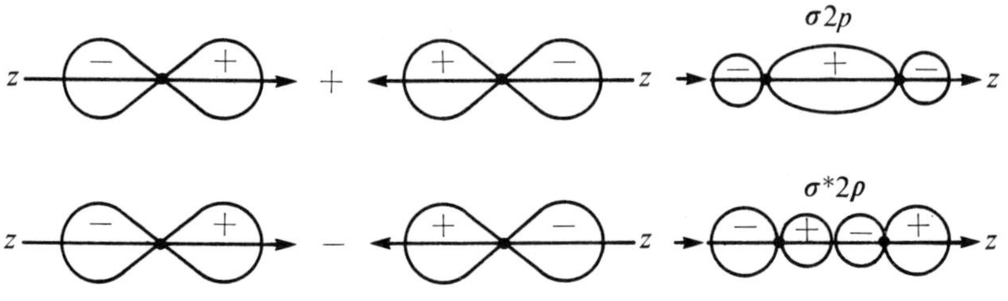

Figure 7.6
Sigma Combinations of p-Orbitals

bining two p orbitals pointing toward each other in a plus and minus linear combination:

$$\phi = a_1\psi^1_{p_z} \pm a_2\psi^2_{p_z}$$

The plus linear combination forms a bonding molecular orbital; the negative combination forms an antibonding molecular orbital because the electron density between the two nuclei is diminished.

Starting with a $2p_y$ orbital from each atom, as depicted in Figure 7.7, we will again construct two molecular orbitals by taking the appropriate linear combinations of the atomic orbitals: $\phi = 1/\sqrt{2}(\psi^1_{p_y} \pm \psi^2_{p_y})$. Be-

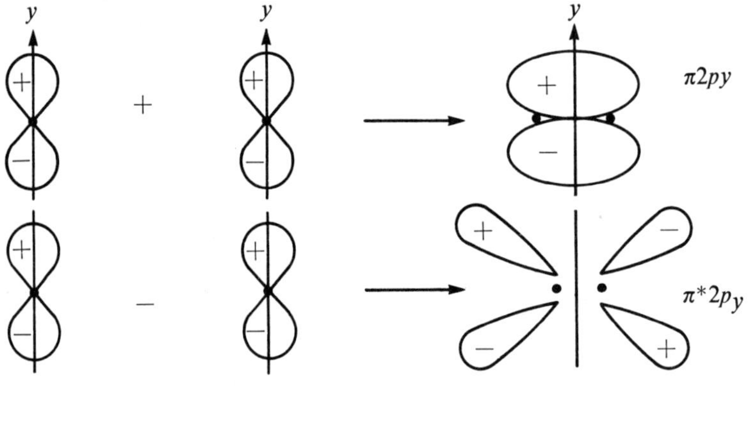

Figure 7.7
P Orbital Pi Combinations

cause of the orientation of the $2p_y$ atomic orbitals to the internuclear axis a π bond is formed. The π bonding molecular orbitals are illustrated in Figure 7.7. Once again one bonding molecular orbital and one antibonding molecular orbital are constructed. The result is the same for the $2p_x$ orbitals.

Now that the bonding and antibonding molecular orbitals are constructed, one of the relative orderings of the one-electron energy levels is shown in Figure 7.8. The energy calculations would be the same for any pair of molecular orbitals (bonding and antibonding) as it was for the 1*s* case, which was illustrated earlier in this chapter. A symmetrical splitting (the bonding level is stabilized by the same amount of energy that the antibonding level is destabilized) results only if no overlap between the two atomic orbitals ($S = 0$) is assumed. As we mentioned earlier, if the overlap between orbitals is included in a more sophisticated calculation, then the energy splittings may no longer be symmetrical. Also, when the 2*s*-2*p* energy separation of the isolated atom is small, then the LCAO for the σ-bonding levels must include both the 2*s* and the $2p_z$ orbital contributions. This result changes the relative orderings of the molecular orbitals' energy levels; for example, the two $\pi 2p$ energy levels may be lower in energy than the $\sigma 2p$ level. Regardless of the specific ordering of the energy levels, the electrons belonging to the diatomic molecule are placed in the

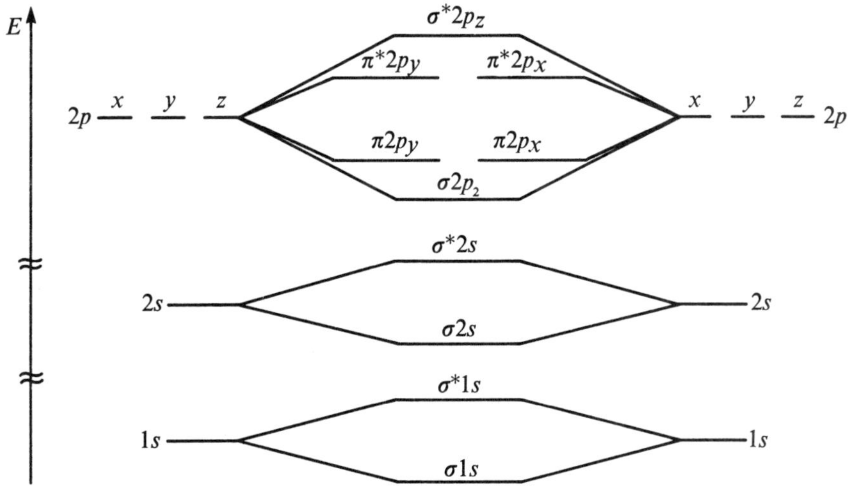

Figure 7.8
One-electron Energy Levels

Table 7.1
Molecular Orbital Electronic Configurations

Molecule		*Configuration*	*Bond Order*
Li_2	$(\sigma 1s)^2(\sigma^* 1s)^2(\sigma 2s)^2$		1
Be_2	$(\sigma 1s)^2(\sigma^* 1s)^2(\sigma 2s)^2(\sigma^* 2s)^2$		0
B_2	*KK*	$(\sigma 2s)^2(\sigma^* 2s)^2(\pi 2p_x)^1(\pi 2p_y)^1$	1
C_2	*KK*	$(\sigma 2s)^2(\sigma^* 2s)^2(\pi 2p_x)^2(\pi 2p_y)^2$	2
N_2	*KK*	$(\sigma 2s)^2(\sigma^* 2s)^2(\pi 2p_{xy})^4(\sigma 2p_z)^2$	3
O_2	*KK*	$(\sigma 2s)^2(\sigma^* 2s)^2(\pi 2p_{xy})^4(\sigma 2p_z)^2(\pi^* 2p_x)^1$ $(\pi^* 2p_y)^1$	2
F_2	*KK*	$(\sigma 2s)^2(\sigma^* 2s)^2(\pi 2p_{xy})^4(\sigma 2p_z)^2(\pi^* 2p_{xy})^4$	1
Ne_2	*KK*	$(\sigma 2s)^2(\sigma^* 2s)^2(\pi 2p_{xy})^4(\sigma 2p_z)^2(\pi^* 2p_{xy})^4$ $(\sigma^* 2p_z)^2$	0

molecular levels following the same rules that apply to the individual atoms. Table 7.1 shows the electron configuration and bond order for all homonuclear diatomic molecules of the second period of the periodic table, based upon the available experimental data. For B_2 and O_2 the $(\pi 2p_x)(\pi 2p_y)$ levels are lower in energy, or more stable, than the $(\sigma 2p_z)$ level, and B_2 is found to be paramagnetic as predicted, while C_2 is diamagnetic.

Exercise 7-1

Show the electron configuration and bond order for Mg_2 and Cl_2.

To conclude this section, the details of the theory for heteronuclear diatomics will be discussed. The first example to be considered, LiH, is a very simple case. The hydrogen atom has only one low energy atomic orbital, and while the lithium atom has five low lying orbitals, the electrons in the lithium 1*s* orbital are not involved in bonding, and the three lithium 2*p* orbitals are not occupied by electrons.

As shown in Figure 7.9, a σ bond can be formed between the hydrogen 1*s* orbital and the lithium 2*s* orbital with some contribution to the molecular orbital from the lithium $2p_z$ orbital. Although the $2p_y$ and $2p_x$ orbitals of the lithium atom have the correct symmetry for π bonding, as shown in Figure 7.9(c) no π bonds are constructed because the 1*s* orbital of the hydrogen atom is incapable of π bonding, and the $2p_x$ and $2p_y$ orbitals of hydrogen are very high in energy.

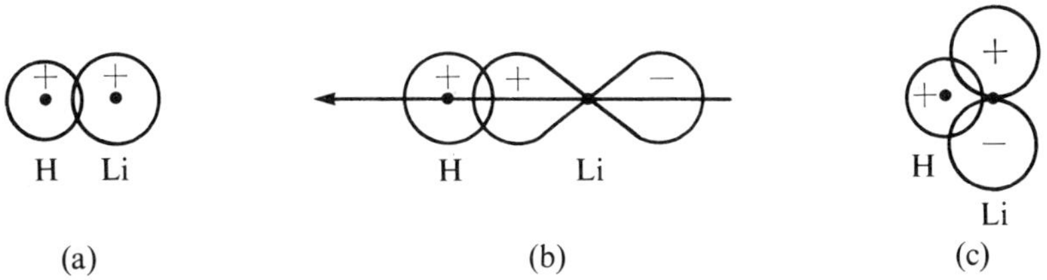

Figure 7.9
Orbital Combinations for LiH

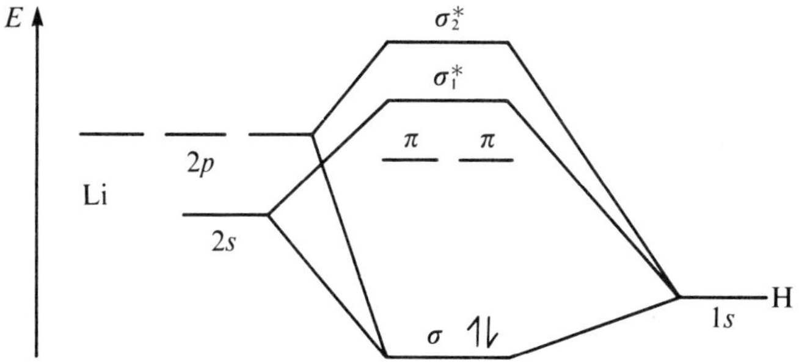

Figure 7.10
LiH *Molecular Orbital Diagram*

In Figure 7.10 we have constructed the qualitative molecular orbital diagram. Since hydrogen is more electronegative than lithium the molecular orbital diagram shows the 2*s* orbital of Li at a higher energy than the 1*s* orbital of H. As we previously mentioned, the σ molecular orbitals are constructed primarily from the H 1*s* and the Li 2*s* orbitals with some contribution from the Li $2p_z$ orbital. A qualitative wave function representation for the bonding molecular orbital might be $\phi_\sigma = c_1\psi_{1s} + c_2\psi_{2s} + c_3\psi_{2p_z}$ where c_3 would be small and $c_1 > c_2$ because the 1*s* orbital of H is more stable than the Li 2*s* and the electrons would be expected to spend more time near the H nucleus than near the Li nucleus. The electronegativities of Li and H would also correspond to this same unequal sharing of the two valence electrons. Because the bonding molecular orbital is constructed using three atomic orbitals, there must be two additional σ molecular orbitals, both of which are antibonding. The qualitative wave functions for these molecular orbitals would be $\phi_{\sigma 1}^* = c_3\psi_{2s} - c_4\psi_{1s}$

and $\phi^*_{\sigma_2} = c_5\psi_{2p_z} - c_6\psi_{1s}$ where $c_3 > c_4$ and $c_5 > c_6$, which shows that both σ^* antibonding levels are centered toward the Li nucleus.

Because the $2p_x$ and $2p_y$ orbitals of Li can not π bond with the $1s$ orbital of H, these atomic orbitals become nonbonding molecular orbitals. Therefore, the two valence electrons for this molecule would be placed in the σ molecular level, the lowest level available, giving for LiH the configuration $(\sigma)^2$. A quantitative treatment of LiH will be discussed later in this chapter.

Heteronuclear diatomic molecules with both atoms in the same period of the periodic table will be discussed now. The energies of atomic orbitals of one atom in the heteronuclear diatomic molecule may be very similar to the energies of the orbitals of the second atom in the molecule, or the energies of the orbitals may be very different. For example, in the former case an AB molecule will have almost the same molecular orbital diagram as the A_2 molecule previously discussed. The only difference would be that the atomic orbitals belonging to the more electronegative element will be slightly more stable, and this will cause the electrons in the molecular orbitals to spend more time near the nucleus of that atom. If B is more electronegative than A, then electrons are attracted toward the B nucleus and the wave function would be $\phi = c_1\psi_A + c_2\psi_B$ with $c_2 > c_1$. Figure 7.11 shows the molecular orbital diagram for the CO molecule, which has 10 electrons in the valence orbitals and 14 electrons altogether. The same

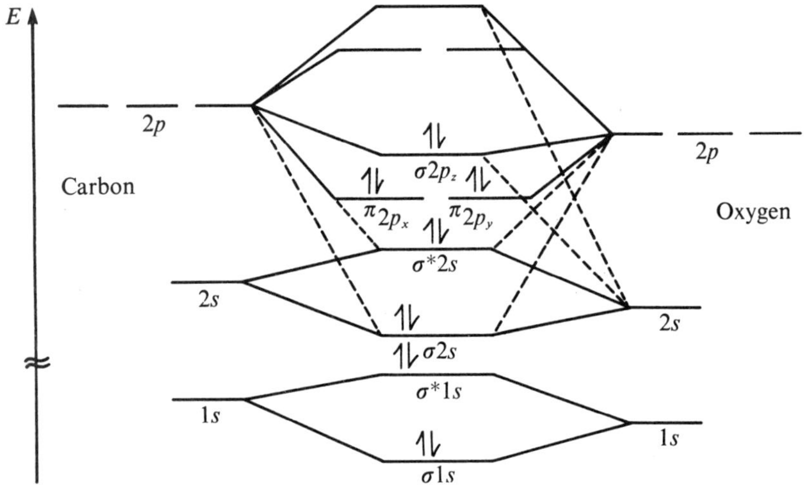

Figure 7.11
CO *Molecular Orbital Diagram*

shorthand notation used for homonuclear diatomic molecules may also be used here; that is, for CO the configuration is

$$(\sigma 1s)^2(\sigma^* 1s)^2(\sigma 2s)^2(\sigma^* 2s)^2(\pi 2p_x)^2 = (\pi 2p_y)^2(\sigma 2p_z)^2$$

The full tie lines to the atomic orbitals in Figure 7.11 represent major contributions to the molecular orbitals and the dashed tie lines, less influential contributions. These latter contributions may cause the $\pi 2p_{xy}$ levels to be more stable than the $\sigma 2p_z$ level. From our diagram and the definition of bond order we can assign a bond order of 3 for CO with one σ bond and two π bonds.

The same procedure applied to the CO molecule may be applied to other heteronuclear diatomic molecules and ions provided the proper atomic orbitals are used in the construction of the linear combinations of atomic orbitals.

Exercise 7-2

Construct a qualitative molecular orbital diagram for the CN^- ion. What is the predicted bond order?

7-3 TRIATOMIC MOLECULES

The simplest triatomic molecule to deal with would be BeH_2 which is linear. In constructing a molecular orbital diagram for this triatomic molecule we will mix the valence atomic orbitals of the central beryllium atom with linear combinations of the valence atomic orbitals of the two outer hydrogen atoms. In this case our molecular orbitals are constructed from the 2*s* and 2*p* orbitals of beryllium and linear combinations of the 1*s* orbitals of hydrogen.

According to the symmetry associated with atomic orbitals only specific linear combinations of atomic orbitals will form bonding molecular orbitals. For example, in Figure 7.12 we can see that the 2*s* orbital of Be can combine with the positive linear combination of hydrogen 1*s* orbitals, $(\psi_{1s_1} + \psi_{1s_2})$. The molecular orbital in Figure 7.12(a) $(\phi = c_1\psi_{2s} + c_2(\psi_{1s_1} + \psi_{1s_2}))$ represents a bonding molecular orbital and that in Figure 7.12(b) an antibonding molecular orbital of the form $(\phi^* = c_3\psi_{2s} - c_4(\psi_{1s_1} + \psi_{1s_2}))$. In Figure 7.13 we can see the possibilities for σ bonding using the $2p_z$ orbital of Be. As previously noted we will designate the σ bonding axis as the *z*-axis, which means that the $2p_x$ and $2p_y$ orbitals of Be are π bonding orbitals but in this case are nonbonding because there are

Figure 7.12
Orbital Combinations for BeH_2

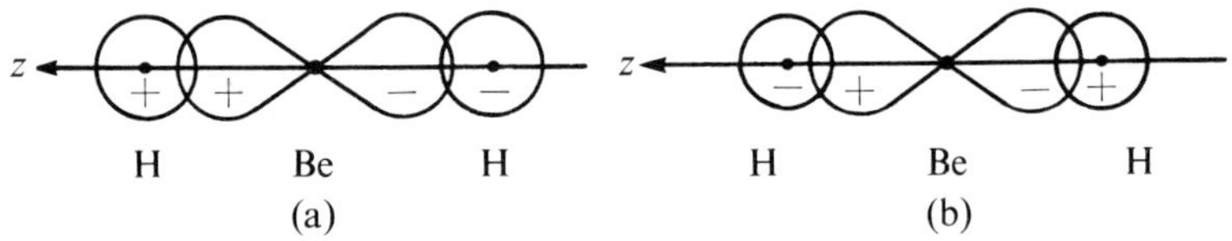

Figure 7.13
Orbital Combinations for BeH_2

no π bonding orbitals on the H atom. The molecular orbital in Figure 7.13(a) ($\phi = c_1\psi_{2p_z} + c_2(\psi_{1s_1} - \psi_{1s_2})$) represents a bonding molecular orbital and the molecular orbital in Figure 7.13(b) an antibonding molecular orbital of the form ($\phi^* = c_3\psi_{2p_z} - c_4(\psi_{1s_1} - \psi_{1s_2})$). The four molecular orbitals in Figures 7.12 and 7.13 represent the possible linear combinations of Be valence orbitals with combinations of H valence orbitals.

Figure 7.14 shows the construction of the molecular orbital energy level diagram for BeH_2 with the valence electrons placed in the appropriate levels. The linear combinations of H orbitals are lower in energy than the Be orbitals because hydrogen is more electronegative than Be. We would expect the electrons in the molecular orbitals to spend more time nearer the hydrogen atoms, or in other words, the bonding molecular orbitals will have more hydrogen character than beryllium character.

The CO_2 molecule is also linear, but because the oxygen atoms are capable of π bonding, linear combinations of π bonding p orbitals must be constructed. The σ bonding interactions are between C 2*s* and O 2*s*, C 2*p* and O 2*s*, C 2*s* and O 2*p*, and C 2*p* and O 2*p*; the first two interactions are similar to the interactions in the BeH_2 case and the third example of bonding in CO_2 involving p_z orbitals is shown in Figure 7.15.

Let us now look along the *x*-coordinate for each atom in the molecule,

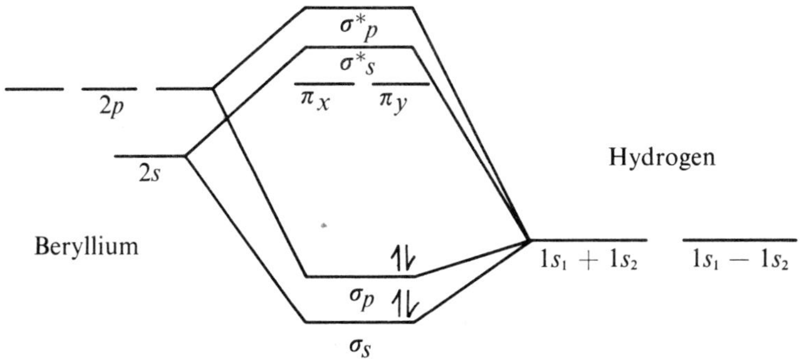

Figure 7.14
BeH$_2$ *Molecular Orbital Diagram*

bonding MO	anti-bonding MO
$(\phi = c_1\psi_{2s} + c_2(\psi_{2p_z^1} + \psi_{2p_z}))$	$(\phi^* = c_3\psi_{2s} - c_4(\psi_{2p_z^1} + \psi_{2p_z}))$

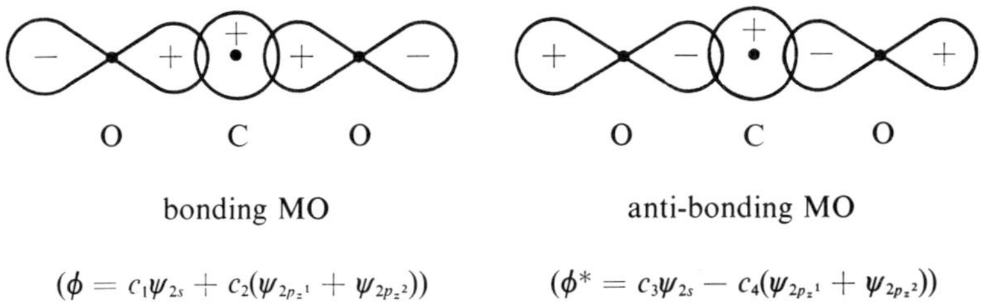

Figure 7.15
P *Orbital Sigma Bonding*

choosing these axes to lie in the plane of the page. The y-coordinate for each atom will be perpendicular to the page. Figure 7.16 illustrates the three types of π bonding molecular orbitals that may be formed. The molecular orbitals illustrated here for the $2p_x$ orbitals would be the same as for the three $2p_y$ orbitals except for the direction. Taking this result and noting the greater electronegativity of oxygen, the molecular orbital energy level diagram for CO_2 is illustrated in Figure 7.17. The $2s$ orbitals of oxygen have not been mixed into the molecular orbital scheme. The CO_2 molecule has 16 valence electrons which are located in the bonding and nonbonding molecular orbitals of the diagram.

The molecular orbital scheme for CO_2 illustrates that as we add more atoms, which means more orbitals, the diagram becomes more complica-

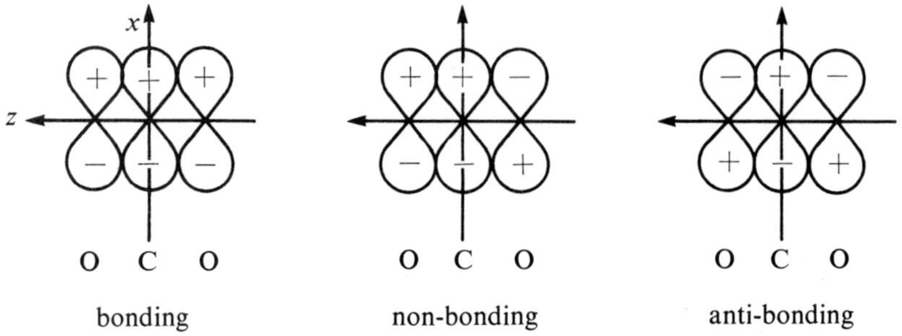

$$\phi_b = c_1\psi_{2p_x} + c_2(\psi_{2p_{x^1}} + \psi_{2p_{x^2}})$$
$$\phi_a^* = c_3\psi_{2p_x} - c_4(\psi_{2p_{x^1}} + \psi_{2p_{x^2}}) \qquad \phi_n = (1/\sqrt{2})(\psi_{2p_{x^1}} - \psi_{2p_{x^2}})$$

Figure 7.16
P Orbital Pi Bonding

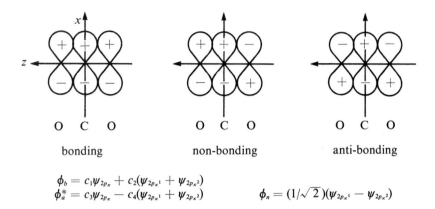

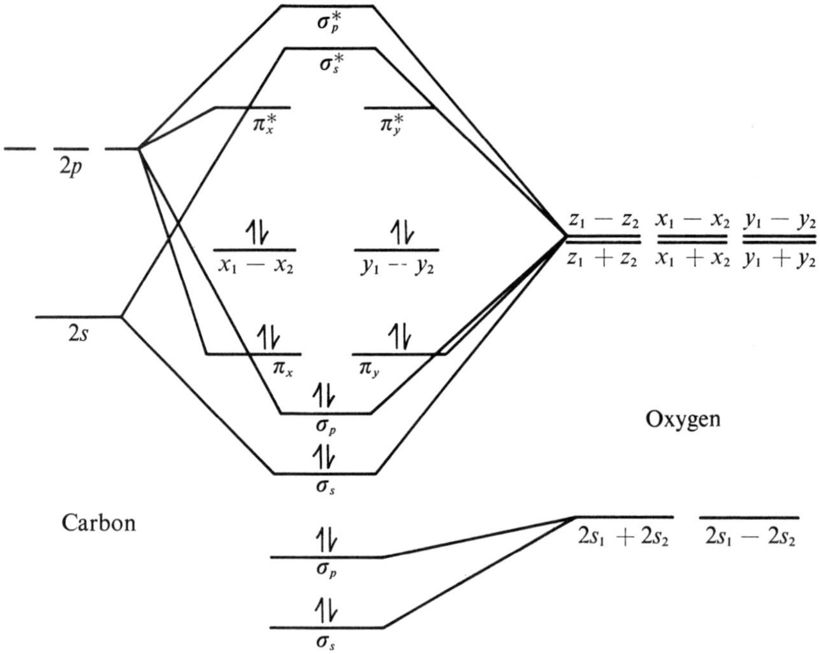

Figure 7.17
CO_2 *Molecular Orbital Diagram*

ted, even when some of the atomic orbitals are not mixed into the linear combinations.

The next type of molecule we will look at is the angular triatomic molecule, H_2O. Because of this angle, two of the $2p$ orbitals of oxygen are capable of σ bonding. As shown by the two bonding molecular orbitals in Figure 7.18, this can make the construction of proper linear combinations even more complicated. Instead of treating the H_2O molecule in the same way as we have treated the other molecules in this chapter, we will deal with this molecule in the next section with the group theoretical techniques discussed in the first two chapters; these techniques will be very useful in generating the linear combinations of atomic orbitals that make up the molecular orbitals. Not only can the proper LCAO's be constructed more easily using group theory but quantitative calculations are more readily discussed. Quantitative calculations may enable us to predict such properties as electronic spectral transitions, especially for the coordination compounds of the first row transition metals.

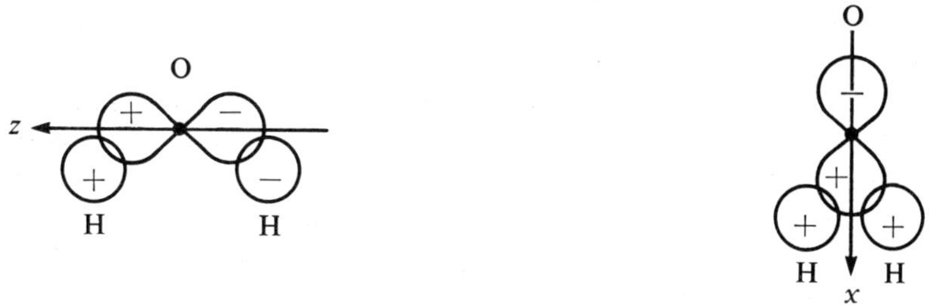

Figure 7.18
Orbital Combinations For H_2O

Exercise 7-3

Construct a qualitative molecular orbital diagram for O_3.

7-4 SELECTED EXAMPLES

The first molecule to be discussed in terms of group theory and molecular orbital theory will be the one with which we ended our previous discussion, H_2O. The symmetry operations for this molecule constitute the C_{2v} point group. The character table for this point group appears in Table 7.2, and is repeated in Appendix I. The valence orbitals of the oxygen atom

Table 7.2
C_{2v} Character Table

C_{2v}	E	C_2	$\sigma_v(xz)$	$\sigma_v'(yz)$		
A_1	1	1	1	1	z	x^2, y^2, z^2
A_2	1	1	−1	−1	R_z	xy
B_1	1	−1	1	−1	x, R_y	xz
B_2	1	−1	−1	1	y, R_x	yz

transform as certain irreducible representations of the C_{2v} point group. In reference to the right-hand columns of the character table, the 2s ($x^2 + y^2 + z^2$) orbital transforms as A_1, the $2p_z$ orbital as A_1, the $2p_x$ orbital as B_1, and the $2p_y$ orbital as B_2. In constructing molecular orbitals earlier in this chapter we had to combine atomic orbitals that had the appropriate symmetry properties. The linear combinations of hydrogen 1s orbitals must transform as A_1 in order to combine with the oxygen 2s and $2p_z$ orbitals, as B_1 to combine with the $2p_x$ orbital, and as B_2 to combine with the $2p_y$ orbital. Let us now set about generating these LCAO's for hydrogen and constructing molecular orbitals using this irreducible representation classification.

Figure 7.19 shows the H_2O molecule and a specific coordinate system for the oxygen atom. The hydrogen atoms require no coordinate system because only the spherically symmetrical 1s orbital is used, but we have numbered these atoms so that they are distinguishable. Table 7.3 shows what happens to each σ bond when each operation of the C_{2v} point group is performed. The reducible representation for the σ bonding in the H_2O molecule can be found by calculating the traces of these representation matrices as we did in Chapter 2, but by referring to Table 7.2 we can arrive at Γ_{red} more easily. For the identity operation, E, the trace of the 2 by 2

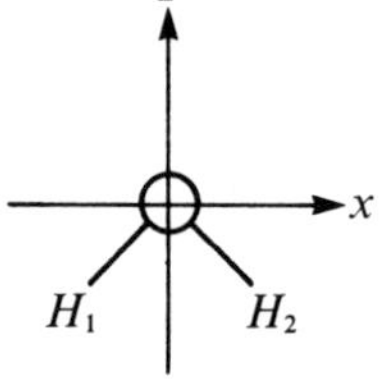

Figure 7.19
The H_2O *Molecule*

Table 7.3

C_{2v} Transformation Table

	E	C_2	$\sigma(xy)$	$\sigma(yz)$
σ_1	σ_1	σ_2	σ_1	σ_2
σ_2	σ_2	σ_1	σ_2	σ_1

representation matrix will equal two, which also corresponds to the number of σ bonds that remain unchanged. The trace of the representation matrix would equal zero for the C_2 operation, because both σ bonds move when this operation is performed. By using this shortcut we get

	E	C_2	$\sigma(xz)$	$\sigma(yz)$
$\Gamma_{red} =$	2	0	2	0

This representation can now be reduced in the usual way to yield the linear combination of irreducible representations: $\Gamma_{red} = A_1 + B_1$. The linear combination of hydrogen 1*s* orbitals that corresponds to the A_1 irreducible representation can be constructed by taking the four bonds designated in Table 7.3 and placing the four entries of the A_1 irreducible representation in that table as coefficients:

$$(+1)\sigma_1 \quad (+1)\sigma_2 \quad (+1)\sigma_1 \quad (+1)\sigma_2$$
$$(+1)\sigma_2 \quad (+1)\sigma_1 \quad (+1)\sigma_2 \quad (+1)\sigma_1$$

The numbers in parentheses are the irreducible representation entries taken from the C_{2v} character table. Upon normalization both rows yield $(1/\sqrt{2})(\sigma_1 + \sigma_2)$. The irreducible representation also contains a B_1 term, and the linear combination of hydrogen σ bonding orbitals corresponding in symmetry to B_1 can be found similarly:

$$(+1)\sigma_1 \quad (-1)\sigma_2 \quad (+1)\sigma_1 \quad (-1)\sigma_2$$
$$(+1)\sigma_2 \quad (-1)\sigma_1 \quad (+1)\sigma_2 \quad (-1)\sigma_1$$

The first row yields the normalized LCAO, $(1/\sqrt{2})(\sigma_1 - \sigma_2)$, while the second row yields $(1/\sqrt{2})(\sigma_2 - \sigma_1)$. Both of these LCAO's show that the negative linear combination of hydrogen 1*s* orbitals bond with the $2p_x$ orbital of oxygen. The result of putting this information in tabular form is shown in Table 7.4. From this table we can see that because the 2*s* and $2p_z$ orbitals of oxygen and the positive linear combination of 1*s* orbitals of hydrogen all have A_1 symmetry, they will mix to form molecular orbitals with A_1 symmetry.

Table 7.4
H_2O Orbital Scheme

Irreducible representation	*Oxygen orbitals*	*Hydrogen's LCAO's*
A_1	$2s$	$(1/\sqrt{2})(\sigma_1 + \sigma_2)$
	$2p_z$	
B_1	$2p_x$	$(1/\sqrt{2})(\sigma_1 - \sigma_2)$
B_2	$2p_y$	———

The molecular orbital energy levels can now be calculated. In Equations (7-3a) and (7-3b) we have constructed the A_1 and B_1 secular equations.

$$A_1 \qquad \begin{matrix} 2s & 2p_z & \psi(H_{1s}) \end{matrix}$$

$$\begin{vmatrix} H_{11} - EG_{11} & H_{12} - EG_{12} & H_{13} - EG_{13} \\ H_{21} - EG_{21} & H_{22} - EG_{22} & H_{23} - EG_{23} \\ H_{31} - EG_{31} & H_{32} - EG_{32} & H_{33} - EG_{33} \end{vmatrix} = 0 \qquad \textbf{(7-3a)}$$

$$B_1 \qquad \begin{matrix} 2p_x & \psi(H_{1s}) \end{matrix}$$

$$\begin{vmatrix} H_{11} - EG_{11} & H_{12} - EG_{12} \\ H_{21} - EG_{21} & H_{22} - EG_{22} \end{vmatrix} = 0 \qquad \textbf{(7-3b)}$$

Any S_{ij} or G_{ij} term where $i = j$ will equal one. In the A_1 determinant $G_{12} = G_{21} = 0$ because the $2s$ and $2p_z$ orbitals of oxygen are orthogonal. The off-diagonal H_{ij} terms (exchange integrals) can be calculated by using an empirical method such as the modified Wolfsberg-Helmholz approximation: $H_{ij} = -FG_{ij}(H_{ii} \times H_{jj})^{1/2}$. H_{ii} and H_{jj} are the coulombic integrals for the atomic orbitals, G_{ij} is the group overlap for orbital i with orbital j, and F is an empirical constant which is most often taken to be 2.0. A group overlap differs from an atomic overlap by accounting for the number of atoms bonded to the central atom and the geometry in which these atoms are arranged. In the H_2O molecule there are two hydrogen atoms bonded to the oxygen atom at an angle of $\sim 105°$. The simplified determinants are shown in Equations (7-4a) and (7-4b), and the solutions

$$A_1 \qquad \begin{matrix} 2s & 2p_z & \psi(H_{1s}) \end{matrix}$$

$$\begin{vmatrix} H_{11} - E & H_{12} & H_{13} - EG_{13} \\ H_{21} & H_{22} - E & H_{23} - EG_{23} \\ H_{31} - EG_{31} & H_{32} - EG_{32} & H_{33} - E \end{vmatrix} = 0 \qquad \textbf{(7-4a)}$$

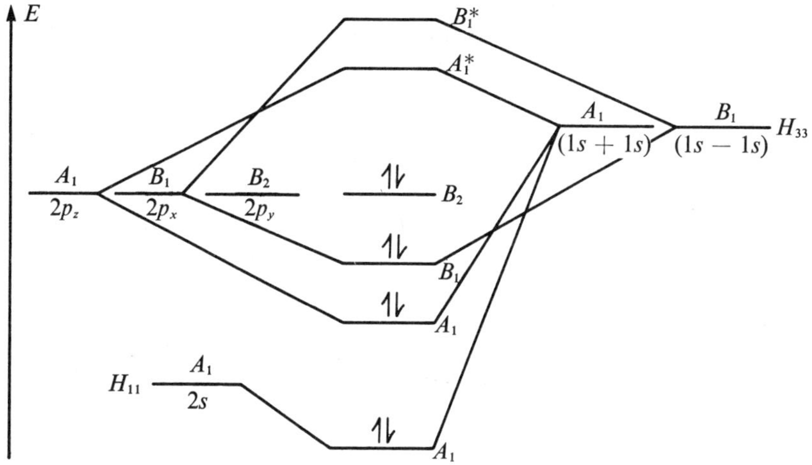

Figure 7.20
H_2O *Molecular Orbital Diagram*

$$\begin{array}{ccc} B_1 & 2p_x & \psi(H_{1s}) \end{array}$$
$$\begin{vmatrix} H_{11} - E & H_{12} - EG_{12} \\ H_{21} - EG_{21} & H_{22} - E \end{vmatrix} = 0 \qquad \textbf{(7-4b)}$$

to these may be found by the normal method. A symmetrical determinant of large dimensions may be easily solved using a computer program. The results of the solution of the A_1 and B_1 determinants is shown in Figure 7.20. The A_1 secular determinant is 3×3 and will have three solutions. From our molecular orbital diagram we can see that two A_1 levels are bonding and one A_1 level is antibonding. The B_1 secular determinant has two solutions, one bonding and one antibonding. The B_2 molecular orbital is nonbonding because no LCAO for hydrogen has B_2 symmetry. The molecular orbital diagram in Figure 7.20 could be constructed quantitatively by choosing numerical values for the H_{ii} and G_{ij} (normalized group overlaps) and by using the modified Wolfsberg-Helmholz approximation during the process of solving for the energies. The oxygen orbitals are lower in energy than the hydrogen orbitals because of the greater electronegativity of oxygen.

Exercise 7-4

Construct the secular equations for NO_2. From your result construct a qualitative molecular orbital diagram.

Before we actually perform a specific quantitative calculation, we should mention some of the important approximations made in the different methods of molecular orbital calculations.

The Huckel Molecular Orbital method is probably the best known and most often discussed method because it is most commonly used in determining molecular orbital energy level diagrams for the π electrons in unsaturated hydrocarbons, although atoms such as nitrogen and oxygen may be included in the system with minor adjustments in the calculation.

In Huckel Molecular Orbital (HMO) theory the coulomb integral (H_{ii}) for a carbon atom is designated as α and the resonance integral (H_{ij}, $i \neq j$) is designated as β for adjacent carbon atoms. The overlap between the p orbitals which are π bonding is neglected ($S = 0$). This important approximation is shown in Equation (7-5).

$$S_{ij} = \int \psi_i \psi_j \, d\tau = \begin{matrix} 1 & i = j \\ 0 & i \neq j \end{matrix} \qquad \textbf{(7-5)}$$

To illustrate the Huckel Molecular Orbital method we will first consider the ethylene molecule, which contains two π electrons, one from each carbon atom. Once again, by using the secular equations and secular determinants explained in Chapter 3, we may solve for the π bonding molecular orbitals in terms of α and β. Equation (7-6) shows the general and specific forms of the equation to be solved. If we divide through by β and let $(\alpha - E)/\beta = x$, the result is

$$\begin{matrix} C_1 & C_2 \end{matrix} \qquad\qquad \begin{matrix} C_1 & C_2 \end{matrix}$$

$$\begin{vmatrix} H_{11} - ES_{11} & H_{12} - ES_{12} \\ H_{21} - ES_{21} & H_{22} - ES_{22} \end{vmatrix} = 0 = \begin{vmatrix} \alpha - E & \beta \\ \beta & \alpha - E \end{vmatrix} \qquad \textbf{(7-6)}$$

$$\begin{vmatrix} x & 1 \\ 1 & x \end{vmatrix} = 0$$

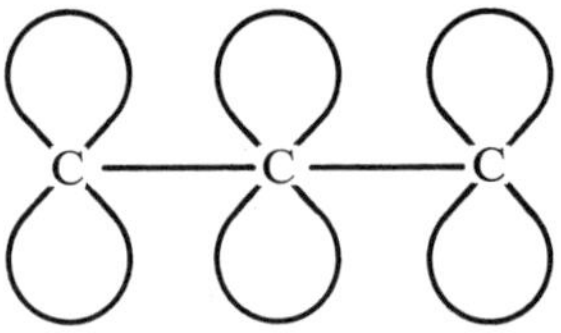

Figure 7.21
The Allyl Radical

Expanding the determinant yields $x^2 - 1 = 0$ and $x = \pm 1$. Substituting $(\alpha - E)/\beta$ back in for x, we find that $E = \alpha + \beta$ and $E = \alpha - \beta$. The results in terms of α and β are comparable to the results we obtained for the H_2 molecule (Figure 7.3) when we assumed no overlap ($S = 0$).

A calculation for the linear three carbon system, shown in Figure 7.21, will further illustrate the principles involved. In the allyl radical each carbon donates one π electron, so we have three electrons and a 3 by 3 determinant as shown in Equation (7-7). We can now solve for the three

$$\begin{array}{ccc} C_1 & C_2 & C_3 \end{array}$$
$$\begin{vmatrix} \alpha - E & \beta & 0 \\ \beta & \alpha - E & \beta \\ 0 & \beta & \alpha - E \end{vmatrix} = 0 \qquad \textbf{(7-7)}$$

π bonding molecular orbitals. Overlap between p orbitals is neglected again and the resonance integral for nonadjacent orbitals is zero. For example, $H_{13} = 0$ because carbon (1) and carbon (3) are not adjacent, but $H_{23} = H_{12} = \beta$ because of adjacent carbon atoms. Dividing through by β and substituting x for $(\alpha - E)/\beta$ yields the determinant shown in Equation (7-8).

$$0 = \begin{vmatrix} x & 1 & 0 \\ 1 & x & 1 \\ 0 & 1 & x \end{vmatrix} \qquad \textbf{(7-8)}$$

Expansion of this determinant in the usual way yields $x^3 - 2x = 0$, or $(x)(x^2 - 2) = 0$. The three roots are $x = 0$, $E = \alpha$; $x = \sqrt{2}$, $E = \alpha - \sqrt{2}\beta$; and $x = -\sqrt{2}$, $E = \alpha + \sqrt{2}\beta$. The resulting molecular orbital energy level diagram is shown in Figure 7.22. For the allyl radical, one electron is in the nonbonding level. For propylene both π electrons would be in the bonding molecular orbital.

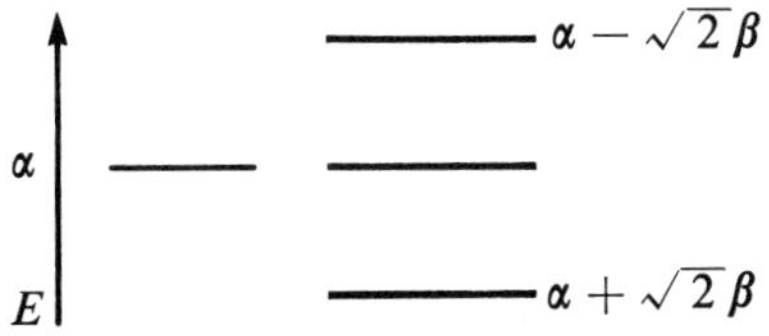

Figure 7.22
Energy Level Diagram for the Allyl Radical

When nitrogen or oxygen atoms are present in the unsaturated molecule, the coulomb integrals are modified as shown in Equation (7-9).

$$\begin{aligned} \alpha_N &= \alpha + \lambda_N \beta \\ \alpha_0 &= \alpha + \lambda_0 \beta \end{aligned} \tag{7-9}$$

When one electron from nitrogen is contributed to the π-system, $\lambda_N = 1/2$; and when two electrons are part of the π-system, $\lambda_N = 3/2$. For oxygen when one electron is used $\lambda_0 = 3/2$ and when two electrons are used $\lambda_0 = 5/2$. The calculations are a little more involved, but the assumptions and principles are the same for all types of Huckel Molecular Orbital calculations.

Exercise 7-5

Using Huckel molecular orbital theory construct the energy levels for butadiene (C_4H_6) and cyclobutadiene (C_4H_4).

There are other approximate molecular orbital theories. The Complete Neglect of Differential Overlap (CNDO) is the most elementary of these. Only the valence electrons of each atom are considered, and the zero differential overlap approximation is used for all products of atomic orbitals whether the atomic orbitals that are overlapping are on different atoms or on the same atom. Overlap integrals are neglected in normalizing the molecular orbitals just as they were for the hydrogen molecule case. Penetration effects, where electrons on one atom forming the bond penetrate the orbitals of the other atom, are also taken into account in determining the coulomb integrals (H_{ii}).

A modification of the CNDO method is the Intermediate Neglect of Differential Overlap (INDO), which takes into account the different interactions that arise between two electrons on the same atom with parallel or antiparallel spins. This modification allows us to distinguish between the states arising from the same electronic configuration. The other approximations involved and further details of these methods will not be discussed here, but can be found in Pople and Beveridge's book.*

LiH

Earlier in this chapter we spoke of a qualitative molecular orbital diagram for the LiH heteronuclear diatomic molecule. We will now follow the same procedures used for the H_2O molecule and construct the molecular orbital

---

* J. A. Pople and D. L. Beveridge, *Approximate Molecular Orbital Theory* (New York: McGraw-Hill Book Co., 1970).

diagram for LiH using secular equations. Equation (7-10) shows the

$$\begin{array}{ccc} 2s & 2p & 1s \end{array}$$

$$\begin{vmatrix} H_{11} - ES_{11} & H_{12} - ES_{12} & H_{13} - ES_{13} \\ H_{21} - ES_{21} & H_{22} - ES_{22} & H_{23} - ES_{23} \\ H_{31} - ES_{31} & H_{32} - ES_{32} & H_{33} - ES_{33} \end{vmatrix} = 0 \qquad \textbf{(7-10)}$$

equation which must be solved. $S_{11} = S_{22} = S_{33} = 1.000$, $S_{12} = S_{21} = 0.000$, and from a paper by Karo and Olson* $S_{13} = S_{31} = 0.469$ and $S_{23} = S_{32} = 0.506$. (The sign of S_{13} was changed to make all overlap values positive.) In order to solve this determinant for the three energy levels we must determine values for the H_{ii}. Let us begin by constructing the Hamiltonian for a one-electron calculation, in order to consider the LiH^+ molecule ion. The Hamiltonian (in atomic units)† is

$$H = -\frac{\nabla^2}{2} - \frac{1}{r'} - \frac{3}{r} + 2\,V(1s^2)$$

where $-\nabla^2/2$ is the kinetic energy
$-1/r'$ is the attractive potential due to the H nucleus
$-3/r$ is the attractive potential due to the Li nucleus

and $2V(1s^2)$ is the repulsion between the bonding electron of the lithium atom and the two nonbonding $1s$ electrons of lithium. In treating the LiH molecule we must add the repulsive potential due to the second electron which is in the σ bonding molecular orbital. However, since we do not know the molecular wave function, we will assume an electron density distribution, add these extra terms to the Hamiltonian, calculate the energies of the molecular orbitals, and then use the secular equations to generate the coefficients of the σ bonding molecular orbital. If the generated coefficients agree with the assumed electron density distribution, then we made an excellent guess. If they do not agree, the next step is to rewrite the Hamiltonian using the newly calculated electron distribution, calculate the energies, and generate the new molecular orbitals. This process is repeated until self-consistency is attained, and is called the SCF method.

Let us assume that the electrons are equally distributed between the two atoms and can be expressed as $1/2\,\phi_{1s}^2 + 1/4\,\phi_{2s}^2 + 1/4\,\phi_{2p}^2$. The Hamiltonian is

$$H = -\frac{\nabla^2}{2} - \frac{1}{r'} - \frac{3}{r} + 2V(1s^2) + \frac{1}{2}V(\phi_{1s}^2) + \frac{1}{4}V(\phi_{2s}^2) + \frac{1}{4}V(\phi_{2p}^2)$$

---

* A. M. Karo and A. R. Olson, *J. Chem. Phys.*, *30*, 1232 (1959).
† See Appendix III

The coulombic integral (H_{11}) for the $2s$ orbital of Li can be calculated by computing the appropriate integrals. One part of the total integral is

$$\int \phi_{2s}\left(-\frac{\nabla^2}{2} - \frac{3}{r} + 2V(1s^2)\right)\phi_{2s}\,d\tau = -.198 \text{ a.u.}$$

The other parts are

$$\int \phi_{2s}\left(\frac{1}{r'}\right)\phi_{2s}\,d\tau = -.264 \qquad \frac{1}{2}\int \phi_{2s}(V(\phi_{1s}^2))\phi_{2s}\,d\tau = .043$$

$$\frac{1}{4}\int \phi_{2s}(V(\phi_{2s}^2))\phi_{2s}\,d\tau = .058 \text{ and } \frac{1}{4}\int \phi_{2s}(V(\phi_{2p}^2))\phi_{2s}\,d\tau = .011$$

The summation of terms shows that $H_{11} = -0.349$. For each H_{ii} evaluated, all terms of the Hamiltonian must be computed. We will not try to evaluate all of these integrals here, but the values may be obtained instead from Karo and Olson's work. A procedure similar to the one outlined here can also be used to calculate H_{22} and H_{33} in Equation (7-11). The exchange integrals may be calculated in the same way. Equation (7-11) shows the secular equation arrived at by assuming the equally distributed

$$\begin{array}{ccc} 2s & 2p & 1s \\ \end{array}$$
$$\begin{vmatrix} -.349 - E & 0 & -.320 - .469E \\ 0 & -.274 - E & -.241 - .506E \\ -.320 - .469E & -.241 - .506E & -.449 - E \end{vmatrix} = 0 \qquad \textbf{(7-11)}$$

electron density. Solution of the secular equation yields the following eigenvalues: $E = -0.499$, -0.293, and $+0.070$. Using the secular equations the normalized wave function for the lowest lying molecular orbital would be

$$\psi_1 = 0.742\,\phi_{1s} + 0.425\,\phi_{2s} - 0.038\,\phi_{2p}$$

To determine the electron density distribution we must square the wave function

$$\psi_1^2 = 0.548\,\phi_{1s}^2 + 0.181\,\phi_{2s}^2 + 0.0014\,\phi_{2p}^2 + 0.631\,\phi_{1s}\phi_{2s} - 0.0561\,\phi_{1s}\phi_{2p}$$

If we equally divide the overlap density between the two atoms, we obtain

$$\phi_{1s}^2 = 0.548 + \tfrac{1}{2}(0.631)(0.469) - \tfrac{1}{2}(0.0561)(0.506) = 0.6838$$
$$\phi_{2s}^2 = 0.265 \qquad\qquad \phi_{2p}^2 = 0.0517$$

This result is not in agreement with the initial assumption of electron density distribution, but we can now rewrite the Hamiltonian with this improved function as shown in Equation (7-12). The process is repeated until the calculated electron densities agree with the assumed electron densities.

$$H = -\frac{\nabla^2}{2} - \frac{1}{r'} - \frac{3}{r} + 2V(1s^2) + 0.684V(\phi_{1s}^2) + 0.265V(\phi_{2s}^2) + .052V(\phi_{2p}^2) \qquad \textbf{(7-12)}$$

The energies found in this final step would be the energies of the molecular orbitals shown earlier in Figure 7.10. The H_{ii} values used in the final step are plotted as the energies of the atomic orbitals before these orbitals were mixed to construct the molecular orbitals.

MnF_6^{3-}

The principles discussed in this chapter can also be applied to transition metal complexes, such as the MnF_6^{3-} unit found in the crystal structure of MnF_3. However, instead of using the self-consistency procedure used for the LiH molecule where we determined all the H_{ii}, we will estimate the H_{ii} values for the fluoride ion from ionization potential data and then use another type of self-consistency procedure to find the H_{ii} for Mn^{3+}.

We will begin the problem by generating the linear combinations of atomic orbitals, using the same method that we used in our discussion of the H_2O molecule. In Figure 7.23 the structure of MnF_6^{3-} is shown with

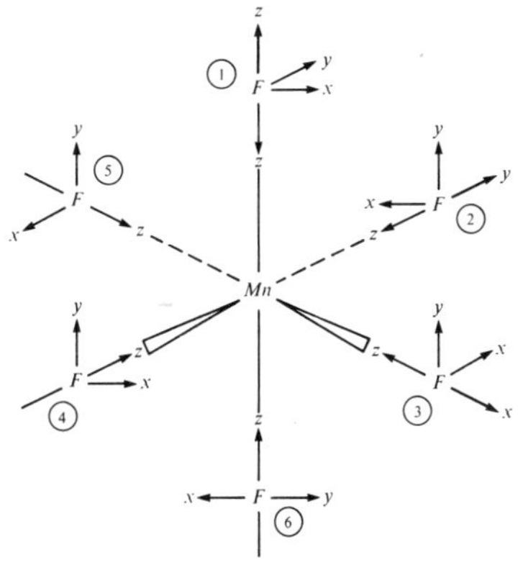

Figure 7.23
Coordinate System for M.O. Calculation for MnF_6^{3-}

coordinate systems assigned to each atom, since the $2p$ orbitals of fluorine are capable of both σ and π bonding. The coordinate system for the Mn^{3+} ion is right-handed and the coordinate systems for the fluoride ions are left-handed. If we specify that all $2p_z$ orbitals of the F^- ions point toward the metal ion, the $2p_x$ and the $2p_y$ orbitals will be available for π bonding. Along each axis the two fluoride ions are the same distance from the manganese ion, but each axis has a different bond distance; this means that the MnF_6^{3-} ion is not octahedral but exhibits a rhombic distortion of the octahedron. Upon investigation of the symmetry elements we find that the MnF_6^{3-} ion belongs to the D_{2h} point group. The D_{2h} character table is found in Appendix I and is repeated for convenience below along with a table for each of the σ bonds in the MnF_6^{3-} ion (Table 7.5). The reducible representation for σ bonding is determined from this listing of bond movements:

	E	$C_2(z)$	$C_2(y)$	$C_2(x)$	i	$\sigma(xy)$	$\sigma(xz)$	$\sigma(yz)$
$\Gamma_{red} =$	6	2	2	2	0	4	4	4

Table 7.5

D_{2h} Character Table

D_{2h}	E	$C_2(z)$	$C_2(y)$	$C_2(x)$	i	$\sigma(xy)$	$\sigma(xz)$	$\sigma(yz)$		
A_g	1	1	1	1	1	1	1	1		x^2, y^2, z^2
B_{1g}	1	1	−1	−1	1	1	−1	−1	R_z	xy
B_{2g}	1	−1	1	−1	1	−1	1	−1	R_y	xz
B_{3g}	1	−1	−1	1	1	−1	−1	1	R_x	yz
A_u	1	1	1	1	−1	−1	−1	−1		
B_{1u}	1	1	−1	−1	−1	−1	1	1	z	
B_{2u}	1	−1	1	−1	−1	1	−1	1	y	
B_{3u}	1	−1	−1	1	−1	1	1	−1	x	

Transformation Table for Sigma Orbitals of MnF_6^{3-} *in* D_{2h}

	E	$C_2(z)$	$C_2(y)$	$C_2(x)$	i	$\sigma(xy)$	$\sigma(xz)$	$\sigma(yz)$
σ_1	σ_1	σ_1	σ_6	σ_6	σ_6	σ_6	σ_1	σ_1
σ_2	σ_2	σ_4	σ_2	σ_4	σ_4	σ_2	σ_4	σ_2
σ_3	σ_3	σ_5	σ_5	σ_3	σ_5	σ_3	σ_3	σ_5
σ_4	σ_4	σ_2	σ_4	σ_2	σ_2	σ_4	σ_2	σ_4
σ_5	σ_5	σ_3	σ_3	σ_5	σ_3	σ_5	σ_5	σ_3
σ_6	σ_6	σ_6	σ_1	σ_1	σ_1	σ_1	σ_6	σ_6

The irreducible representation is found to be $\Gamma_{\text{red}} = 3A_g + B_{1u} + B_{2u} + B_{3u}$. Multiplying each entry in the transformation table by the corresponding character of the irreducible representation we obtain the following σ bonding linear combinations for the fluorine orbitals:

$$\begin{array}{llll} 3A_g & 4(\sigma_1 + \sigma_6), & 4(\sigma_2 + \sigma_4), & 4(\sigma_3 + \sigma_5) \\ B_{1u} & 4(\sigma_1 - \sigma_6) \sim 4(\sigma_6 - \sigma_1) & & \\ B_{2u} & 4(\sigma_2 - \sigma_4) \sim 4(\sigma_4 - \sigma_2) & & \\ B_{3u} & 4(\sigma_3 - \sigma_5) \sim 4(\sigma_5 - \sigma_3) & & \end{array}$$

Coefficients are included for the generated LCAO's but they will be absorbed upon normalization of the functions. Referring to the coordinate system diagram in Figure 7.23, we can readily see that these generated linear combinations of fluorine orbitals will have the proper symmetry to overlap with the corresponding metal orbitals. For example, the $4p_z$ orbital of manganese has B_{1u} symmetry and $(1/\sqrt{2})(\sigma_1 - \sigma_6)$ has B_{1u} symmetry. The normalized fluorine LCAO's used for σ bonding are shown in Equation (7-13).

$$\begin{array}{ll} A_g: & (1/\sqrt{2})(\sigma_1 + \sigma_6) \quad (1/\sqrt{2})(\sigma_2 + \sigma_4) \quad (1/\sqrt{2})(\sigma_3 + \sigma_5) \\ B_{1u}: & (1/\sqrt{2})(\sigma_1 - \sigma_6) \\ B_{2u}: & (1/\sqrt{2})(\sigma_2 - \sigma_4) \\ B_{3u}: & (1/\sqrt{2})(\sigma_3 - \sigma_5) \end{array} \tag{7-13}$$

We will now consider the π bonding linear combinations of fluorine orbitals. The first step is to construct a transformation table which will show what happens to each $2p_y$ and $2p_x$ orbital of the fluorines as each operation is performed. The fluorine π-bonding transformation table is constructed by referring to the coordinate systems in Figure 7.23 and is shown in Table 7.6.

The reducible representation for π bonding can be taken directly from Table 7.6.

	E	$C_2(z)$	$C_2(y)$	$C_2(x)$	i	$\sigma(xy)$	$\sigma(xz)$	$\sigma(yz)$
$\Gamma_{\text{red}} =$	12	−4	−4	−4	0	0	0	0

The irreducible representations are found to be $\Gamma_{\text{red}} = 2B_{1g} + 2B_{2g} + 2B_{3g} + 2B_{1u} + 2B_{2u} + 2B_{3u}$. To obtain the linear combinations of fluorine orbitals used for π bonding we again multiply each entry in the transformation table by the corresponding character of the irreducible representation. Including the coefficients obtained, we find the following

Table 7.6
Transformation Table for π-Orbitals

	E	$C_2(z)$	$C_2(y)$	$C_2(x)$	i	$\sigma(xy)$	$\sigma(xz)$	$\sigma(yz)$
x_1	x_1	$-x_1$	x_6	$-x_6$	x_6	$-x_6$	x_1	$-x_1$
x_2	x_2	x_4	$-x_2$	$-x_4$	x_4	x_2	$-x_4$	$-x_2$
x_3	x_3	x_5	$-x_5$	$-x_3$	x_5	x_3	$-x_3$	$-x_5$
x_4	x_4	x_2	$-x_4$	$-x_2$	x_2	x_4	$-x_2$	$-x_4$
x_5	x_5	x_3	$-x_3$	$-x_5$	x_3	x_5	$-x_5$	$-x_3$
x_6	x_6	$-x_6$	x_1	$-x_1$	x_1	$-x_1$	x_6	$-x_6$
y_1	y_1	$-y_1$	y_6	$-y_6$	$-y_6$	y_6	$-y_1$	y_1
y_2	y_2	y_4	$-y_2$	$-y_4$	$-y_4$	$-y_2$	y_4	y_2
y_3	y_3	y_5	$-y_5$	$-y_3$	$-y_5$	$-y_3$	y_3	y_5
y_4	y_4	y_2	$-y_4$	$-y_2$	$-y_2$	$-y_4$	y_2	y_4
y_5	y_5	y_3	$-y_3$	$-y_5$	$-y_3$	$-y_5$	y_5	y_3
y_6	y_6	$-y_6$	y_1	$-y_1$	$-y_1$	y_1	$-y_6$	y_6

combinations (Equation (7-14)):

$$
\begin{array}{ll}
2B_{1g} & 4(x_2 + x_4),\quad 4(x_3 + x_5) \\
2B_{2g} & 4(x_1 + x_6),\quad 4(y_3 - y_5) \sim 4(y_5 - y_3) \\
2B_{3g} & 4(y_1 - y_6) \sim 4(y_6 - y_1),\quad 4(y_2 - y_4) \sim 4(y_4 - y_2) \\
2B_{1u} & 4(y_2 + y_4),\quad 4(y_3 + y_5) \\
2B_{2u} & 4(x_3 - x_5) \sim 4(x_5 - x_3),\quad 4(y_1 + y_6) \\
2B_{3u} & 4(x_1 - x_6) \sim 4(x_6 - x_1),\quad 4(x_2 - x_4) \sim 4(x_4 - x_2)
\end{array}
\qquad (7\text{-}14)
$$

The normalized linear combinations for π bonding are shown in Equation (7-15). We now have the LCAO's for fluoride ions which have the proper

$$
\begin{array}{lll}
B_{1g}: & (1/\sqrt{2})(x_2 + x_4) & (1/\sqrt{2})(x_3 + x_5) \\
B_{2g}: & (1/\sqrt{2})(x_1 + x_6) & (1/\sqrt{2})(y_3 - y_5) \\
B_{3g}: & (1/\sqrt{2})(y_1 - y_6) & (1/\sqrt{2})(y_2 - y_4) \\
B_{1u}: & (1/\sqrt{2})(y_2 + y_4) & (1/\sqrt{2})(y_3 + y_5) \\
B_{2u}: & (1/\sqrt{2})(x_3 - x_5) & (1/\sqrt{2})(y_1 + y_6) \\
B_{3u}: & (1/\sqrt{2})(x_1 - x_6) & (1/\sqrt{2})(x_2 - x_4)
\end{array}
\qquad (7\text{-}15)
$$

Table 7.7
Ligand and Metal Bonding Orbitals

Representation	*Mn*	*Ligand Orbitals*
A_g	$4s$	$(1/\sqrt{2})(\sigma_1 + \sigma_6)$; $(1/\sqrt{2})(\sigma_2 + \sigma_4)$; $(1/\sqrt{2})(\sigma_3 + \sigma_5)$
	$3d_{x^2-y^2}$	
	$3d_{z^2}$	
B_{3u}	$4p_x$	$(1/\sqrt{2})(\sigma_3 - \sigma_5)$; $(1/\sqrt{2})(x_1 - x_6)$; $(1/\sqrt{2})(x_2 - x_4)$
B_{2u}	$4p_y$	$(1/\sqrt{2})(\sigma_2 - \sigma_4)$; $(1/\sqrt{2})(x_3 - x_5)$; $(1/\sqrt{2})(y_1 + y_6)$
B_{1u}	$4p_z$	$(1/\sqrt{2})(\sigma_1 - \sigma_6)$; $(1/\sqrt{2})(y_2 + y_4)$ $(1/\sqrt{2})(y_3 + y_5)$
B_{1g}	$3d_{xy}$	$(1/\sqrt{2})(x_2 + x_4)$; $(1/\sqrt{2})(x_3 + x_5)$
B_{2g}	$3d_{xz}$	$(1/\sqrt{2})(x_1 + x_6)$; $(1/\sqrt{2})(y_3 - y_5)$
B_{3g}	$3d_{yz}$	$(1/\sqrt{2})(y_1 - y_6)$; $(1/\sqrt{2})(y_2 - y_4)$

symmetry to overlap and form bonds with the $4s$, $4p$, and $3d$ orbitals of manganese. The results are summarized in Table 7.7 in the D_{2h} metal-ligand orbital scheme. For the A_g ligand orbitals linear combinations of the generated LCAO's have been chosen in order to provide for all of the possible σ bonding with the metal ion.

The next step is to construct a secular equation for each irreducible representation, but before we do this let us choose values for the overlap integrals (S_{ij}) and the coulomb integrals (H_{ii}). The atomic overlap integrals can be calculated from Slater-type or other appropriate wave functions. Because each axis in our molecule deals with a different Mn—F bond length, there will be three overlap values for each metal-ligand pair of orbitals. An atomic overlap deals with only two atoms and a group overlap deals with the Mn^{3+} ion and the appropriate number of fluoride ions. With six fluorine atoms bonding in this complex ion we must evaluate group overlap values from the atomic overlaps. The most direct method of calculating group overlaps is by use of formulas developed by Kettle* or Yeranos†. The coulomb energies of the atomic orbitals are estimated from spectroscopic data and ionization potentials, and the secular determinants now may be solved. During the course of the diagonalization of

* S. F. A. Kettle, *Inorg. Chem.*, *4*, 1821 (1965).
† W. Yeranos, *Inorg. Chem.*, *5*, 2070 (1966).

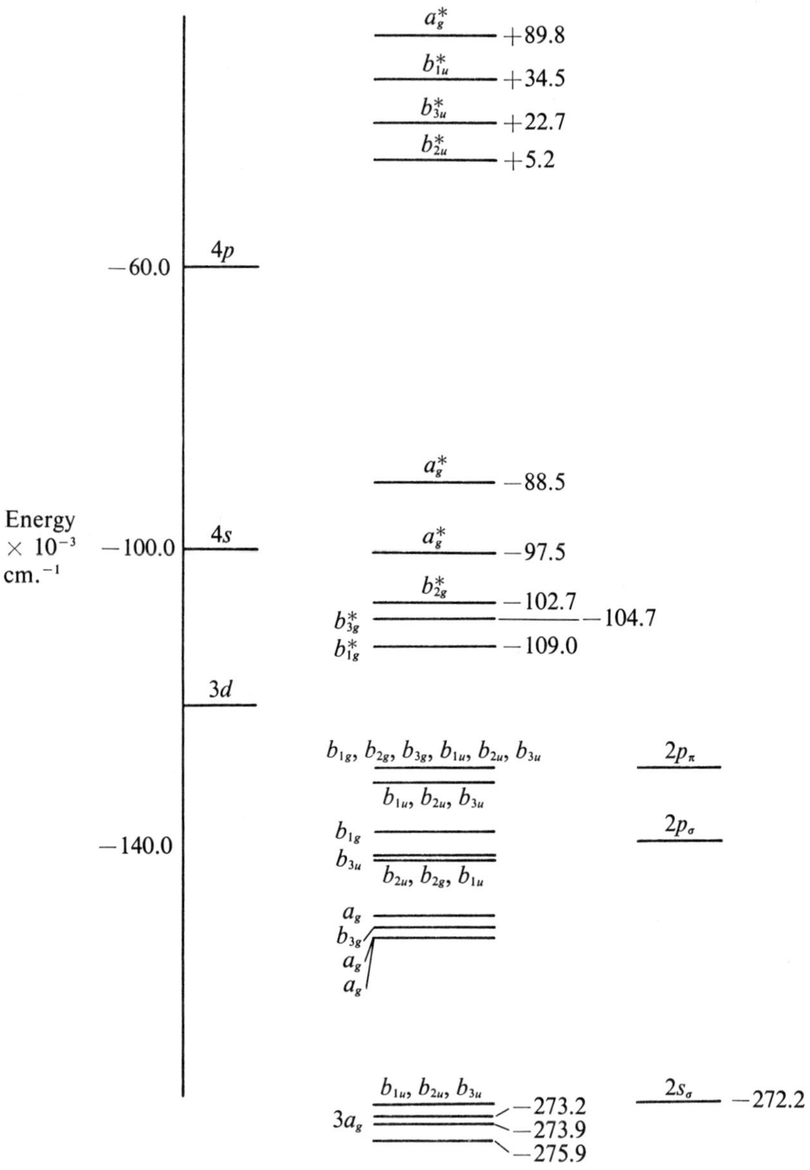

Figure 7.24
Molecular Orbital Energy Level Diagram for MnF_6^{3-}

each determinant the off-diagonal resonance integrals are calculated using the Wolfsberg-Helmholz approximation: $H_{ij} = -FG_{ij}(H_{ii} \times H_{jj})^{1/2}$.

The final molecular orbital energy level diagram for MnF_6^{3-} appears in Figure 7.24. The atomic orbitals of the fluoride ion are lower in energy than the atomic orbitals of the Mn^{3+} ion because the fluorine atom is more electronegative. The $2p_z$ orbital of fluorine, which is the σ bonding orbital, is slightly more stable than the $2p_y$, $2p_x$ degenerate π bonding pair of orbitals. For the metal ion the $3d$ atomic orbitals are lower in energy than the $4s$ orbital which in turn is lower than the $4p$ atomic orbitals. The electrons present in these valence atomic orbitals (8 electrons from each F^- and 4 electrons from Mn^{3+}) are placed in the molecular levels. The four most energetic electrons are found in antibonding molecular orbitals that are primarily $3d$ in character. Magnetic measurements show that these four electrons are all unpaired, which follows from the small splitting observed between the five antibonding levels in the molecular orbital diagram. The electronic transitions found for these levels would be $a_g^* \leftrightarrow a_g^*$, $b_{2g}^* \leftrightarrow a_g^*$, $b_{3g}^* \leftrightarrow a_g^*$, and $b_{1g}^* \leftrightarrow a_g^*$. Since the b_{1g}^*, b_{2g}^*, and b_{3g}^* energy levels have similar energy, the three most energetic transitions appear as one broad band with shoulders. The spectrum of MnF_3 is shown in Figure 7.25, where the vertical lines represent calculated transitions. The apparent agreement between theory and experiment is not bad, if we consider the abundant approximations and the neglect of interelectronic repulsion.

In this chapter we have shown a few examples of how molecular orbital theory can be applied to different types of molecules. The ability of molecular orbital theory to permit quantitative calculations of molecular properties may be questionable, but the formalism of this theory provides a

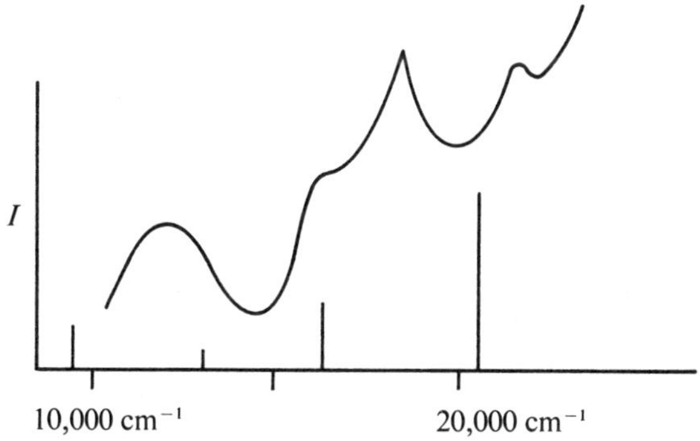

Figure 7.25
Spectrum of MnF_6^{3-} *Ion in* MnF_3

convenient method for the systematization of many kinds of experimental data.

Exercise 7-6

Construct the sigma-bonding and pi-bonding LCAO's for square planar $PtCl_4^{2-}$.

7-5 ORBITAL SYMMETRY IN REACTION MECHANISMS*

In previous sections of this chapter we have discussed bonding and antibonding molecular orbitals and have shown how they are constructed using semiempirical theory and symmetry techniques. The same principles of symmetry and the concept of net positive overlap may be applied to the discussion of reaction mechanisms, where it is assumed that the activated complex is a molecule which is formed from the reactants. The symmetry techniques are similar to those that we have worked with previously, the main difference being the overlap of molecular orbitals from different molecules to form the activated complex rather than just the overlap of atomic orbitals to form molecular orbitals.

During the course of the reaction the most important molecular orbitals are the highest occupied molecular orbital (HOMO) of one reactant and the lowest unoccupied molecular orbital (LUMO) of the other reactant. When the activated complex is formed, electrons flow from the HOMO to the LUMO. This process can take place only if there is net positive overlap between the two interacting molecular orbitals. The HOMO and LUMO must also be close in energy (within about 6 *eV*) or the flowing electrons would not be able to make the transfer. The smaller this energy difference the more easily the reaction may take place. The flow of electrons from the HOMO to the LUMO must also correspond to the breaking and forming of bonds that lead to the desired products.

The reaction $H_2 + I_2 \longrightarrow 2HI$ will serve as a good first example. In Figure 7.26 we see the molecular orbital diagrams formed from the valence orbitals of the reactants. In order to break the H—H bond the electrons must leave the HOMO of H_2 and flow into the LUMO of I_2. Since the LUMO of I_2 is an antibonding orbital, the addition of the two electrons from H_2 leads to the breaking of the I—I bond. In Figure 7.27 the *concerted* addition of H_2 to I_2 via a $\sigma_s(H_2) \longrightarrow \pi_p^*(I_2)$ process is shown to have no net overlap at the activated complex "molecular" stage and hence the process

* See, for example, W. B. Pearson, *Chem. and Engr. News*, September 28, 1970, pp. 66 ff.

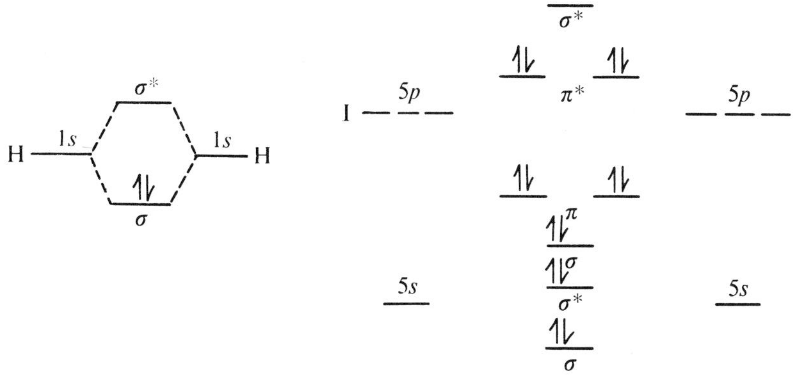

Figure 7.26
H_2 *and* I_2 *Molecular Orbital Diagrams*

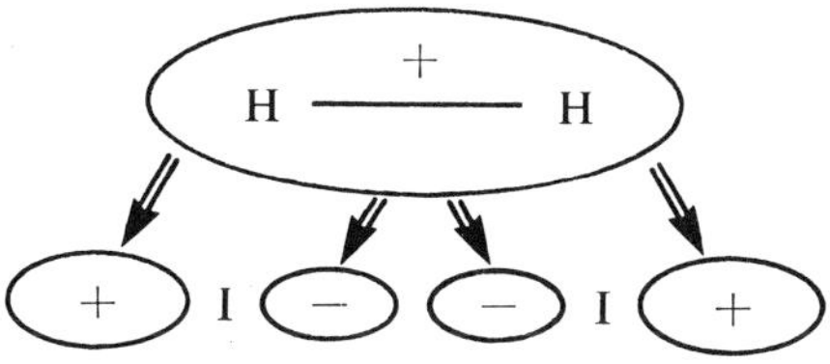

Figure 7.27
Electron Flow from H_2 *to* I_2

is said to be *symmetry forbidden.* The molecular orbitals represented here have been constructed by methods described in Section 7-2. Sullivan first proposed that reaction by broadside collision of the two molecules was forbidden by symmetry in 1967. Until then it was the accepted mechanism for this reaction.

If we now look at the opposite direction of electron flow, where the HOMO of I_2 is π^* and the LUMO of H_2 is σ^*, the mechanism is symmetry allowed as shown in Figure 7.28 but makes no sense chemically because the I—I bond is strengthened instead of weakened by the loss of electrons from an antibonding orbital. This mechanism also contradicts what we have learned about electronegativity since electrons should flow toward the more electronegative atom and of course that is iodine.

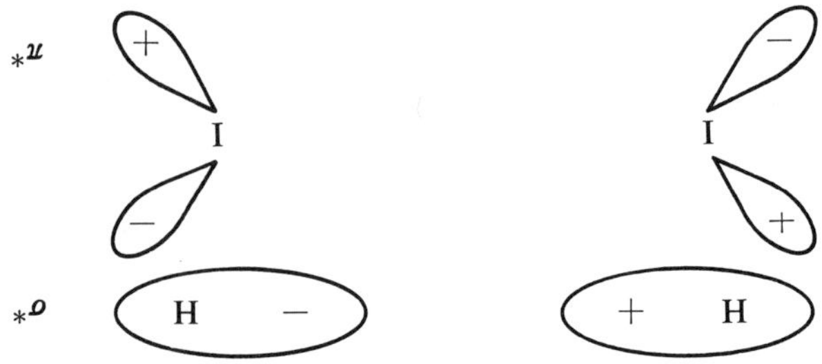

Figure 7.28
Electron Flow from I_2 *to* H_2

Exercise 7-7

Show that either the mechanism $2I + H_2 \longrightarrow 2\,HI$ or $I + H_2 \longrightarrow HI + H$ is symmetry allowed.

We shall now consider the addition of an appropriate reagent to a double bond. The reaction to be discussed is the addition of H_2 to ethylene, i.e. $H_2 + C_2H_4 \longrightarrow C_2H_6$. This reaction can be looked at in two ways. If the electrons flow from H_2 to C_2H_4 the molecular orbitals appear as shown in Figure 7.29, where it may be seen that the mechanism is symmetry forbidden. If the electrons flow in the opposite direction as pictured in Figure 7.30, the reaction is also symmetry forbidden. A simple one-step mechanism for this reaction is not likely. However, one four addition of H_2 and

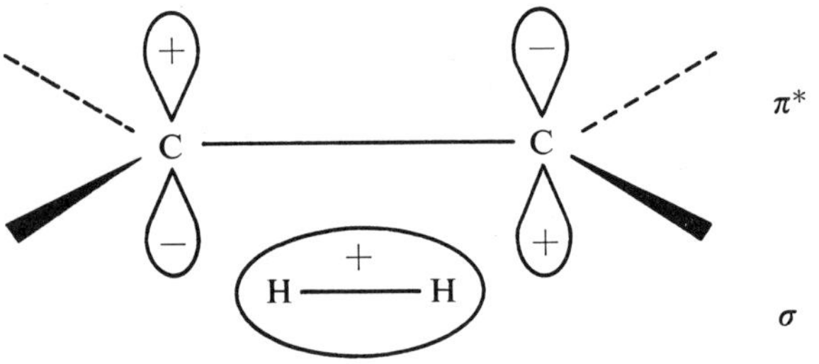

Figure 7.29

Electron Flow from H_2 *to* $C{=}C$

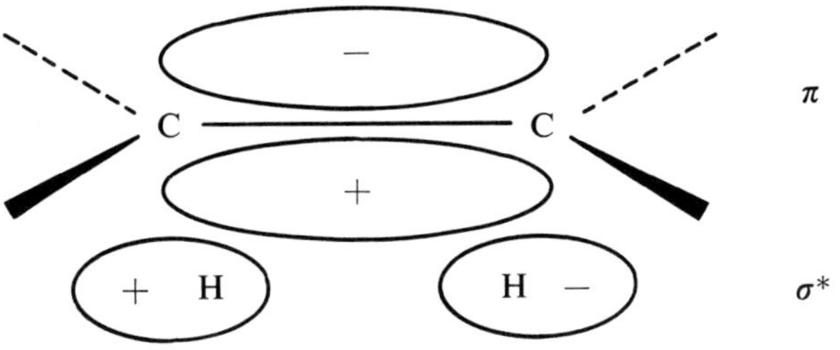

Figure 7.30

Electron Flow from $>C=C<$ *to* H_2

similar reagents to butadiene is symmetry allowed because the π-orbitals of the diene have the proper symmetry to overlap with the antibonding orbitals of the adding reagent.

Exercise 7-8

Using Huckel Molecular Orbital theory show that $CH_2=CH-CH=CH_2$ has the proper molecular orbitals to accept the addition of Cl_2.

For completeness let us examine the addition of a non-symmetric diatomic molecule such as HCl to a double bond. As shown in Figure 7.31,

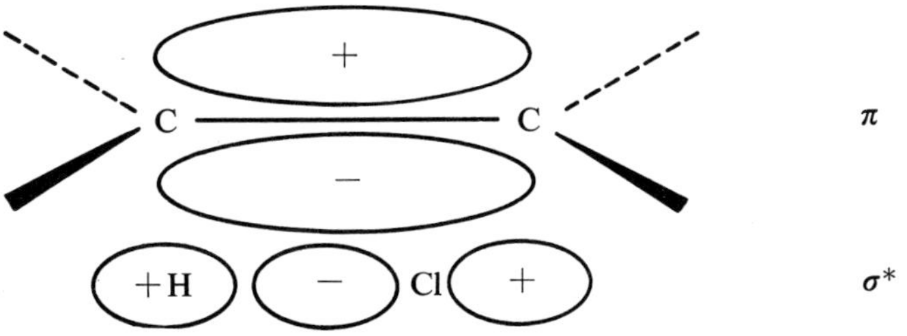

Figure 7.31

Electron Flow from $>C=C<$ *to* HCl

the overlap in this case is not identically zero, but very small. If the addition takes place according to this mechanism, we would expect the activation energy to be quite large. Since the net positive overlap for this bimolecular addition is very small, the mechanism for this reaction most likely involves a stepwise addition. The arguments made here can be extended to any addition reactions involving *XY* molecules where *X* and *Y* are connected by a single bond.

Molecules of H_2 can be added to double bonds with the use of a catalyst. For example, nickel is often used to absorb the H_2 molecules and as suggested in Figure 7.32, the hydrogen molecule first adds to a *d* orbital of the nickel atom that has the correct symmetry. When electrons flow from the filled *d* orbital to the σ^* molecular orbital of H_2, the hydrogen molecule dissociates. The hydrogen atoms are then transferred to the olefin. This mechanism for hydrogenation is symmetry allowed, but may or may not represent the actual process. However, whatever the mechanism may be, each step must correspond to a symmetry-allowed reaction. Some catalyzed reactions are very slow, which may correspond to cases where the reaction is symmetry-forbidden. As we have shown here, the *d* orbitals of the transition metals offer a pathway for the addition of small molecules to olefins.

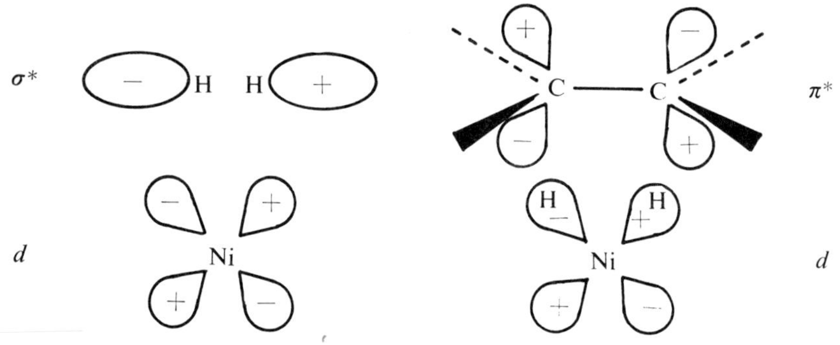

Figure 7.32
Catalytic Addition of H_2

Exercise 7-9

Show that reaction $H_2C{=}CH_2 + O_2 \longrightarrow$ $\begin{array}{c}H_2C{-}CH_2\\ |\quad\ \ |\\ O{-}O\end{array}$ is symmetry forbidden even if O_2 is in its first excited state.

The symmetry aspects of overlapping orbitals in reaction mechanisms are very important in predicting the outcome and mechanism of reactions. The examples discussed here have dealt with simple molecules, but these principles of symmetry can be applied to complex problems of reaction mechanisms and stereochemistry. For a discussion of this topic, the monograph by Woodward and Hoffman* is recommended.

Bibliography

There is much active research in molecular orbital theory. Useful monographs and textbooks include:

C.J. Ballhausen and H.B. Gray, *Molecular Orbital Theory*. New York: W.A. Benjamin, 1965.

H.B. Gray, *Electrons and Chemical Bonding*. New York: W.A. Benjamin, 1965.

J.A. Pople and D.L. Beveridge, *Approximate Molecular Orbital Theory*. New York: McGraw-Hill Book Co., 1970.

L. Salem, *The Molecular Orbital Theory of Conjugated Systems*, New York: W.A. Benjamin, 1966.

A. Streitweiser, *Molecular Orbital Theory for Organic Chemists*, New York: John Wiley and Sons, 1959.

* R. B. Woodward, and R. Hoffman, *The Conservation of Orbital Symmetry* (New York: Academic Press, 1970).

8

SYMMETRY OF THE CRYSTALLINE STATE

An appreciation of the symmetry properties of the crystalline state is necessary in order to interpret the results of many experiments and in order to design many others. In this chapter we provide an elementary discussion of the symmetry properties of the crystalline state. This discussion will include the symmetry elements and operations that lead to space groups, as well as the notation which is encountered most frequently when reading about these symmetry properties. The utility of solid state symmetry concepts will be illustrated by application to the problem which arises in the interpretation of the vibrational spectra of solids.

8-1 HERMANN-MAUGUIN SYMBOLISM

Symmetry elements and operations in the earlier sections of this book were designated by the Schoenflies system, which is usually adopted by those interested in the symmetry properties of isolated molecules. However, those interested in the properties of condensed phases most frequently use an alternate system which is called the Hermann-Mauguin or International notation. For rotation axes, reflection planes, the center of inversion, and identity, a simple relationship exists between the symmetry elements, operations, and symbols of the two systems; these are summarized in Table 8.1. It may be seen from the information collected in the table that rotation axes which are designated by the symbol C_n in the Schoenflies notation are designated by the Arabic numeral which corresponds to the order of the rotation. Reflection planes are given the symbol m; we shall see later how the various mirror planes are differentiated. The inversion center and operation are designated by $\bar{1}$, and the identity is designated by 1.

Table 8.1

Correspondence of Schoenflies and Hermann-Mauguin Symmetry Symbols

Element	*Operation*	*Schoenflies*	*Hermann-Mauguin*
rotation axis	rotation by 360°/n	C_n	n
reflection plane	reflection	σ	m
inversion center	inversion	i	$\bar{1}$
identity	rotation by 360°	E	1

The improper rotation axes are defined differently in the Schoenflies system than in the Hermann-Mauguin system. In the Schoenflies notation we have seen that the improper rotation, S_n, is a rotation followed by a reflection through a mirror plane perpendicular to the rotation axis. In the Hermann-Mauguin system the improper rotation is a rotation of 360°/n followed by inversion through a point on the rotation axis. The symbol is $\bar{n}$.

It is instructive to examine the operations as defined in the two systems. The diagrams in Figure 8.1 show schematically up to order six the results of both the roto-reflection operation and the roto-inversion operation. Here we start with a point which is either above or below the plane of the page as designated by the full or open circle, respectively, carry out the rotation as designated by the order of the axis, and then reflect through the plane in the case of $\tilde{n}$, or invert through the center in the case of $\bar{n}$.

We shall consider in detail the $\bar{3}$ operation in Figure 8.1. Beginning with the point 1 above the plane, we rotate by 120° and then invert through the center encountering point 2 which is below the plane. Continuing the operations as indicated by the serial labeling we find that the $\bar{3}$ improper axis generates the six points as shown. It is a good idea to practice with the other examples shown in the figure in order to gain familiarity with the operation.

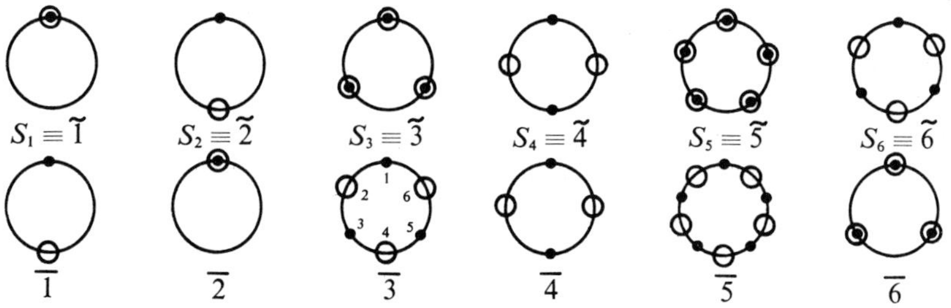

Figure 8.1

Comparison of the Roto-Reflection Operations to the Roto-Inversion Operation: Here ● Designates a Point above the Plane of the Page and ○ Designates a Point below the Plane.

Now let us compare the two sets of symmetry elements and operations and determine which ones are equivalent. Examination of the results in Figure 8.1 reveals the following equivalencies:

$$
\begin{aligned}
\bar{1} &= \tilde{2} \\
\bar{2} &= \tilde{1} \\
\bar{3} &= \tilde{6} \\
\bar{4} &= \tilde{4} \\
\bar{5} &= \widetilde{10} \quad \text{(although not illustrated)} \\
\bar{6} &= \tilde{3}
\end{aligned}
\tag{8-1}
$$

In all cases where n is odd, $\bar{n} = S_{2n}$. Also note that $\bar{1} = S_2 = i$, and $\bar{2} = S_1 = \sigma$; the symbols $\bar{1}$ and m are always used.

8-2 UNIT CELLS

A crystal is made up of atoms or ions which are arranged in a pattern that is repeated periodically in three dimensions. Within this pattern it is possible to choose a set of points, all of which are identical with respect to the repeating pattern. If these points are connected by straight parallel lines, it may be considered that a lattice is constructed. The lattice is made up of a great number of identical parallelepipeds which are related one to the other by translation from one set of lattice points to an equivalent set. In other words, the lattice can be generated by translation of one of the parallelepipeds that was constructed. This generating parallelepiped is called a unit cell. It should be clear that there are numerous ways to choose a unit cell for any given lattice; however, the choice of the unit cell is usually governed by symmetry considerations. The unit cell is specified by the lengths of the three independent edges, a, b, and c, and by the three angles between these edges, α, β, and γ as shown in Figure 8.2.

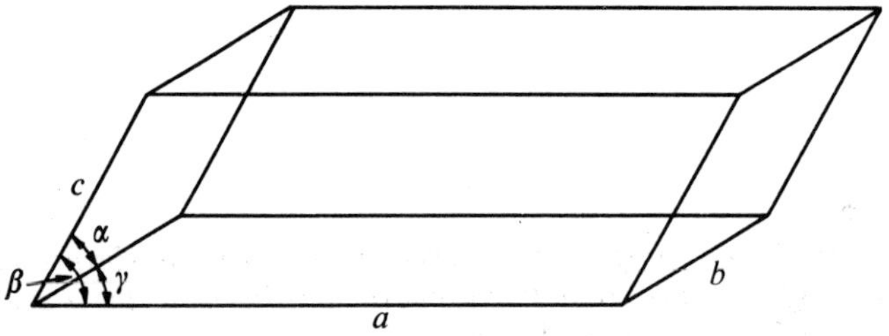

Figure 8.2
Axes and Angles of a Unit Cell

There are seven coordinate systems defined by a, b, c and α, β, γ that are convenient for the description of crystal structures. These coordinate systems are the basis for the broad classification of crystal structures, and the classifications are referred to as the seven crystal systems. These crystal systems are listed in Table 8.2 along with the unit cell parameters. For

Table 8.2
The Seven Crystal Systems

System	*Parameters*
triclinic	$a \neq b \neq c;\ \alpha \neq \beta \neq \gamma \neq 90°$
monoclinic	$a \neq b \neq c;\ \alpha = \gamma = 90°, \beta \neq 90°$
orthorhombic	$a \neq b \neq c;\ \alpha = \beta = \gamma = 90°$
tetragonal	$a = b \neq c;\ \alpha = \beta = \gamma = 90°$
rhombohedral	$a = b = c;\ \alpha = \beta = \gamma \neq 90°$
hexagonal	$a = b \neq c;\ \alpha = \beta = 90°, \gamma = 120°$
cubic	$a = b = c;\ \alpha = \beta = \gamma = 90°$

reasons that will be given later the rhombohedral system is sometimes taken to be a derivative of the hexagonal system, and if that is understood, then there are only six crystal systems.

Any point within the unit cell may be unambiguously specified by three fractional coordinates *x*, *y*, *and z*. To reach the point specified by these coordinates we start at the origin (0, 0, 0) and move along *a* the distance *ax*, then in the *ab* plane we move parallel to *b* the distance *by*, and finally we move the distance *cz* parallel to *c*.

The volume of the unit cell is given by the formula in Equation (8-2),

$$V = abc(1 - \cos^2 \alpha - \cos^2 \beta - \cos^2 \gamma + 2\cos \alpha \cos \beta \cos \gamma)^{1/2} \quad \textbf{(8-2)}$$

and the distance between any two points, say x_1, y_1, z_1, and x_2, y_2, z_2 is given by

$$\begin{aligned} l = [&(x_1 - x_2)^2a^2 + (y_1 - y_2)^2b^2 + (z_1 - z_2)^2c^2 \\ &+ 2(x_1 - x_2)(y_1 - y_2)ab \cos \gamma + 2(y_1 - y_2)(z_1 - z_2)bc \cos \alpha \\ &+ 2(z_1 - z_2)(x_1 - x_2)ca \cos \beta]^{1/2}. \end{aligned} \quad \textbf{(8-3)}$$

Equation (8-3) is very useful since it is frequently necessary to calculate from crystal structure data the distance between two atoms in a unit cell.

Exercise 8-1

In the report of the structure of hexaamminechromium(III) pentachlorocuprate(II), $[Cr(NH_3)_6][CuCl_5]$, the following representative atomic parameters are given for equatorial and axial chloride ions of the trigonal bipyramidal complex anion:

	x	*y*	*z*
Cl (axial)	0.1904	0.1904	0.1904
Cl (equatorial)	0.2500	0.0760	0.0760

Are these two chloride ions coordinated to the same copper ion in a given complex anion? The compound crystallizes in a cubic system with $a = 22.24$A.

8-3 SYMMETRY RESTRICTIONS

Since we are going to let symmetry considerations govern the choice of the unit cell, it is important that we examine specific symmetry elements and operations to see if there are any restrictions, say on the order of the

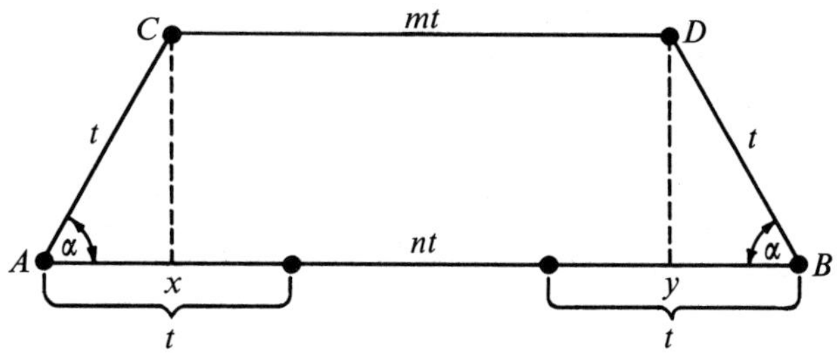

Figure 8.3
Construction Illustrating Symmetry Restrictions on the Order of the Rotation Axis

rotation axes. Consider two lattice points A and B separated by nt, some integral multiple of the unit cell translation distance, as shown in Figure 8.3. From the figure we see that

$$AX = t \cos \alpha$$

and that

$$CD = XY$$

Therefore,

$$XY = AB - AX - BY$$

and by substitution,

$$\begin{aligned} XY &= nt - t \cos \alpha - t \cos \alpha \\ &= t(n - 2 \cos \alpha) \end{aligned}$$

and

$$CD = t(n - 2 \cos \alpha) = mt$$

where

$$m = n - 2 \cos \alpha$$

and n and m are integers. Thus

$$\begin{aligned} \cos \alpha &= (n - m)/2 \\ &= M/2 \end{aligned}$$

where, of course, M is an integer, but more importantly for this discussion we see that cos α can assume only half-integer values. Since $\cos(2\pi/n)$ must lie between -1 and $+1$, the only permitted values are

$$\begin{aligned}
\cos(2\pi/n) &= -1; \quad \text{therefore } n = 2, \text{ and } \alpha = 180^\circ \\
\cos(2\pi/n) &= -.5; \quad \text{therefore } n = 3, \text{ and } \alpha = 120^\circ \\
\cos(2\pi/n) &= 0; \quad \text{therefore } n = 4, \text{ and } \alpha = 90^\circ \\
\cos(2\pi/n) &= .5; \quad \text{therefore } n = 6, \text{ and } \alpha = 60^\circ \\
\cos(2\pi/n) &= 1; \quad \text{therefore } n = 1, \text{ and } \alpha = 360^\circ
\end{aligned} \tag{8-4}$$

Thus, orders of rotation axes are restricted to 1, 2, 3, 4, and 6. It follows that rotatory-inversion axes must also be limited to the same orders.

The possible combinations of rotation axes are also restricted. From one of the basic definitions of group theory it is required that if point 1 is taken to point 2 by operation R_α and then to point 3 by the operation R_β, there exists a third operation R_γ, which is also a member of the group and which will take point 1 to point 3 in one step. The allowed combinations of rotation axes are shown in Equation (8-5).

$$\begin{array}{cccc}
 & R_\alpha & R_\beta & R_\gamma \\
n = & 2 & 2 & 1 \\
 & 2 & 2 & 2 \\
 & 2 & 2 & 3 \\
 & 2 & 2 & 4 \\
 & 2 & 2 & 6 \\
 & 2 & 3 & 3 \\
 & 2 & 3 & 4
\end{array} \tag{8-5}$$

Symmetry restrictions also apply to the combinations of proper rotations with improper rotations. If R_α and R_β are proper rotations, then R_γ must be proper. And, if any two of the rotations are improper, then the third rotation must be proper. There are no other allowed combinations.

8-4 CRYSTALLOGRAPHIC POINT GROUPS

The crystallographic point groups arise from the allowed combinations of the proper and improper rotations mentioned in the last section. We shall see that there are only 32 allowed combinations and, therefore, there are 32 crystallographic point groups. First consider the monoaxial point groups. It is easy to see for the 5 allowed values for the orders of the proper and improper rotations that ten point groups are formed; these are

	crystallographic point groups		
order of rotation	*proper*	*improper*	
1	1	$\bar{1}$	
2	2	$\bar{2} = m$	
3	3	$\bar{3}$	**(8-6)**
4	4	$\bar{4}$	
6	6	$\bar{6}$	

Reference to Figure 8.1 will reveal that the point group $\bar{6}$ is equivalent to a group containing a three fold rotation axis and a mirror plane perpendicular to that axis. In the Hermann-Mauguin system this becomes $3/m$, which is read "three upon m", where $/m$ denotes a plane perpendicular to the three fold axis. In order to generate the remaining monoaxial point groups a mirror plane is taken to be perpendicular to the rotation axis. The results are

$$\begin{aligned} &1/m = m = \bar{2} \\ &2/m \\ &3/m = \bar{6} \\ &4/m \\ &6/m \end{aligned} \tag{8-7}$$

thereby generating three new point groups for a total of 13 monoaxial point groups. The same results can be obtained by combining a roto-inversion axis with a colinear rotation axis. Inspection of the diagrams in Figure 8.1 will serve to illustrate this point.

Using the results in Equation (8-5) we can now write the polyaxial point groups. The results are summarized in Table 8.3, where all the point groups are classified as to the crystal system and as to the corresponding Schoenflies symbols. There are some conventions that are followed in writing the Hermann-Mauguin symmetry symbols. These may be summarized in the following way:

1. Each element in the symmetry symbol for the point group refers to a different direction, and the terms $2/m$, $4/m$, and $6/m$ are single components and refer to only one direction.
2. The position of m in the symbol reflects the direction of the normal to the mirror plane.

3. In the orthorhombic system the three directions are mutually perpendicular. The symbols *mm*2, *m*2*m*, and 2*mm* are all equivalent since they arise from the renaming of the axes.
4. In the tetragonal system 4 or $\bar{4}$ is always taken parallel to the *z* axis. The second place in the symbol refers to the mutually perpendicular *x* and *y* axes, while the third place refers to the bisector of the *x* and *y* axes in the *xy* plane.
5. In the trigonal and hexagonal systems the second place in the symbol refers to equivalent directions in the plane perpendicular to the 3, $\bar{3}$, 6, or $\bar{6}$ axis. An element in the third place refers to bisectors of these equivalent directions.
6. A 3 in the second place always indicates a cubic system. The first place refers to the axes of the cube and the third place to the face diagonals.

Table 8.3

The 32 Crystallographic Point Groups

Crystal System	*Hermann-Mauguin*	*Schoenflies*
Triclinic	1	C_1
	$\bar{1}$	C_i
Monoclinic	2	C_2
	m	C_s
	2/*m*	C_{2h}
Orthorhombic	222	D_2
	*mm*2	C_{2v}
	mmm	D_{2h}
Tetragonal	4	C_4
	$\bar{4}$	S_4
	4/*m*	C_{4h}
	422	D_4
	4*mm*	C_{4v}
	$\bar{4}$2*m*	D_{2d}
	4/*mmm*	D_{4h}
Rhombohedral or Trigonal	3	C_3
	$\bar{3}$	S_6
	32	D_3
	3*m*	C_{3v}
	$\bar{3}$*m*	D_{3d}

Table 8.3 *(continued)*

Crystal System	*Hermann-Mauguin*	*Schoenflies*
Hexagonal	6	C_6
	$\bar{6}$	C_{3h}
	$6/m$	C_{6h}
	622	D_6
	$6mm$	C_{6v}
	$\bar{6}m2$	D_{3h}
	$6/mmm$	D_{6h}
Cubic	23	T
	$m3$ $(\overline{33}2)$	T_h
	432	O
	$\bar{4}3m$	T_d
	$m3m$ $(4\overline{32})$	O_h

8-5 LATTICE SYMMETRY FOR THE CRYSTAL SYSTEMS

Since lattice points are identical they must lie on centers of inversion, and in addition, points midway between lattice points must also be centers of inversion. These inversion centers are the only symmetry elements possessed by the triclinic lattice, and therefore we may assign the symmetry symbol $\bar{1}$ to the triclinic lattice.

The monoclinic unit cell has a mirror plane perpendicular to the *b* axis, and this mirror plane bisects the unit cell; it is also parallel to the *ac* plane. In addition, inspection of Figure 8.4a will show that there is a two-fold rotation axis parallel with the *b* axis. Thus, the monoclinic lattice has symmetry $2/m$.

The orthorhombic unit cell has three mutually perpendicular mirror planes and is given the symbol *mmm*. When the lattice is generated by the orthorhombic unit cell, the faces of the unit cells also become mirror planes of symmetry. The presence of the three mutually perpendicular mirror planes guarantees that additional symmetry elements such as these will be present.

The tetragonal lattice has a four-fold rotation axis, which, by convention, is taken to be parallel to the *c*-axis. There is a mirror plane perpendicular to the four-fold axis, and in addition two unique mirror planes, which are parallel to the *c*-axis and lie 45° apart; other mirror planes may be generated by the four-fold rotation operation. These two unique mirror planes are illustrated in Figure 8.4b, where it may be seen that one mirror

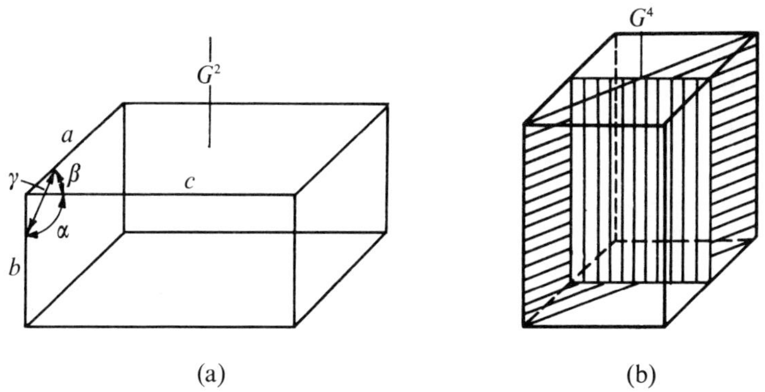

Figure 8.4
(a) Monoclinic Unit Cell (b) Tetragonal Unit Cell

plane bisects the unit cell along the body diagonal while the second plane bisects the edges. These elements of symmetry are indicated by the symbol $4/mmm$.

With these basic ideas in mind it is now possible to see the symmetry of the remaining lattices. The symmetry symbols are

rhombohedral	$\bar{3}m$
hexagonal	$6/mmm$
cubic	$m3m$

8-6 BRAVAIS LATTICES

The seven lattices for the crystal systems listed in Table 8.2 each have lattice points at the corners of the unit cells only, and such lattices are called primitive. Note that there is the equivalent of one lattice point per unit cell in each case. Primitive lattices are designated by the letter P (or R in the case of rhombohedral lattices), which is placed in front of the symmetry symbol for the point group. Thus we have $P\bar{1}$, $P2/m$, $Pmmm$, $P4/mmm$, $R\bar{3}m$, $P6/mmm$, and $Pm3m$.

As first noted by Bravais in 1848, there are some complex lattices which contain more than one lattice point per unit cell, but which still conform to the characteristics of one of the seven crystal systems given in Table 8.2. These nonprimitive lattices may have lattice points in only one of the three sets of parallel faces of the unit cell, or may have one lattice point in the

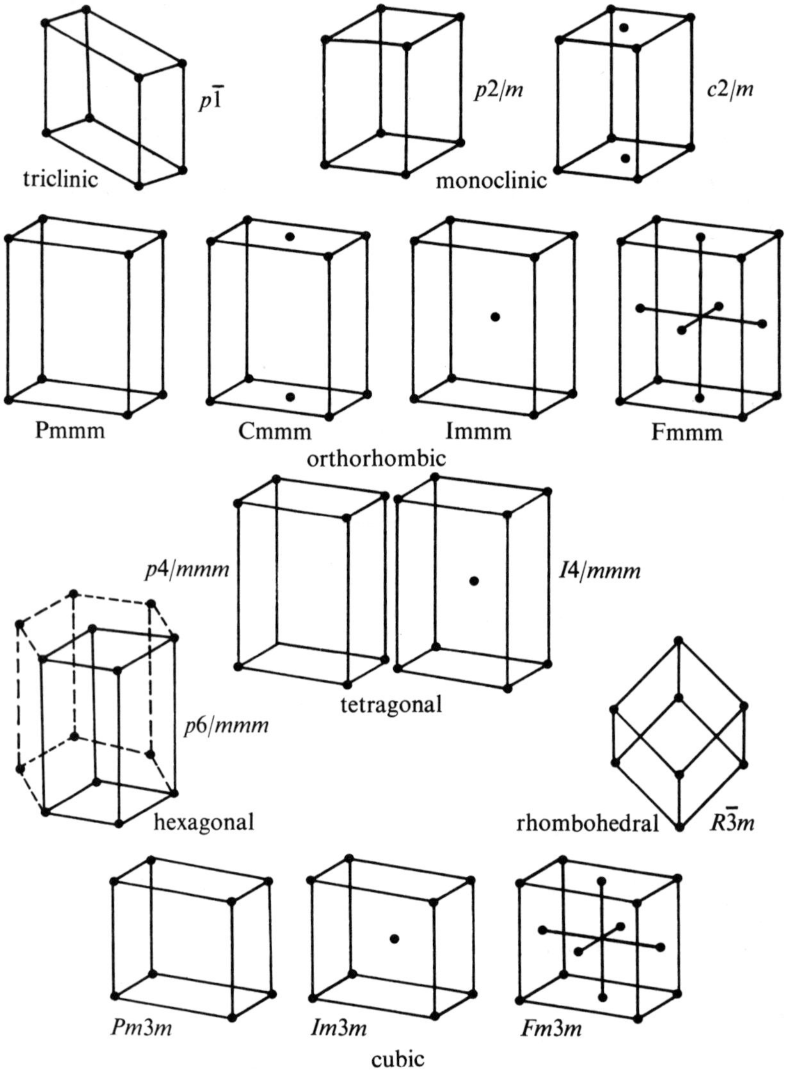

Figure 8.5
The Fourteen Bravais Lattices

center of the unit cell, or may have lattice points in all six faces of the unit cell. If one set of faces is centered, the lattices are designated by *A*, *B*, or *C*, depending on which set of parallel faces are centered. The letter *I* (inner) is used for body-centered lattices, and *F* for face-centered lattices, where all six faces are included.

At this stage there are some points which must be given. By convention a *C*-centered unit cell is usually selected in the case of a monoclinic lattice. Also, a *B*-centered unit cell is never chosen for a monoclinic lattice since in this case it is possible to select a primitive monoclinic unit cell of half the size. Neither are *C*-centered tetragonal unit cells chosen, since it is possible to choose a smaller tetragonal unit cell.

It turns out that there are seven nonprimitive lattices. These, along with seven primitive lattices, give the fourteen Bravais lattices, which are illustrated in Figure 8.5.

It is convenient at this point to discuss the interrelationships between hexagonal, trigonal, and rhombohedral lattices. If there is a 6 or $\bar{6}$ axis in the point group, the crystal is hexagonal, and a primitive cell with $a = b$, $\alpha = \beta = 90°$, and $\gamma = 120°$ may be chosen. This unit cell is illustrated in Figure 8.5. However, if the highest order axis in the point group is 3 or $\bar{3}$, the system is trigonal, but there are two possible lattices. If it is possible to choose a primitive unit cell such that $a = b$, $\alpha = \beta = 90°$, and $\gamma = 120°$, then the lattice is identical with the hexagonal system. It may not be possible to select a primitive cell with these restrictions on dimensions, and if such is the case, it is always possible to select a primitive cell with $a = b = c$, and $\alpha = \beta = \gamma$ which are not necessarily not equal to 90°.

Exercise 8-2

By means of a diagram show how it is possible to choose a smaller unit cell for a *B*-centered monoclinic lattice which is still monoclinic. To which crystallographic point group does this smaller unit cell belong?

8-7 TRANSLATIONAL SYMMETRY OPERATIONS

In repeating patterns there are symmetry operations which are translational in nature. In addition to the operation of translation, by which the lattices are generated from the unit cells, there exist *glide planes* and *screw axes*. A glide plane is a reflection across a mirror plane followed by a translation. If the operation, which is termed a glide, is parallel to the *a* axis, then it is designated by an *a* and consists of reflection through a specified plane followed by a translation of $a/2$. Similar situations may be described for the other two axes. In addition to axial glides, there are face-diagonal glides, which are given the symbol n and involve translation by $(a + b)/2$, $(a + c)/2$, or $(b + c)/2$. The last type of glide is a diamond glide, which is possible only for space groups with face- or body-centered unit cells. Here the translation is $(a + b)/4$, $(a + c)/4$, or $(b + c)/4$.

Let us examine a few examples of glide planes. Consider a b glide perpendicular to c. This means that the mirror plane is perpendicular to the c axis, and that the translation is parallel to the b axis. If we take the point specified by the three coordinates (x, y, z), the reflection in the mirror plane perpendicular to c gives the new position $(x, y, \bar{z})$, and then translation along b gives the symmetry equivalent position $(x, y + \frac{1}{2}, \bar{z})$. A second example should suffice to reinforce the concept of the operation. This time we will take the c glide perpendicular to b. The equivalent positions are

$$(x, y, z) \xrightarrow{m} (x, \bar{y}, z) \xrightarrow{c \text{ translation}} (x, \bar{y}, z + \tfrac{1}{2})$$

As in the case of roto-inversion and roto-reflection axes, the glide symmetry operation is a combination of two separate movements, neither of which need be a valid symmetry operation.

It may be useful to point out that a b glide perpendicular to b is not allowed because the translation is in a direction parallel to the mirror plane. Also note that in monoclinic cells there can be glide planes perpendicular to b only. A c glide perpendicular to a is not possible since a and c are not mutually perpendicular.

We will now turn our attention to the screw axes. The operation accompanying a screw axis is a rotation by $360°/n$ followed by a translation in the direction of the axis. The symbol is n_p and the fraction p/n is the extent of the translation. Thus, the screw axis 2_1 indicates rotation by 180° and translation of one-half the unit cell length parallel to the screw axis. It then follows that a 3_1 screw axis involves rotation by 120° and translation by 1/3 the unit cell length. The permitted screw axes are 2_1, 3_1, 3_2, 4_1, 4_2, 4_3, 6_1, 6_2, 6_3, 6_4, and 6_5. The 2_1 and 4_3 screw axes are illustrated in Figure 8.6. The 4_3 screw axis gives a right handed screw and the 4_1 screw axis gives rise to a left handed screw.

Exercise 8-3

By means of diagrams such as those used in Figure 8.6 show that the 3_1 screw axis gives a right handed screw and that the 3_2 screw axis gives a left handed screw.

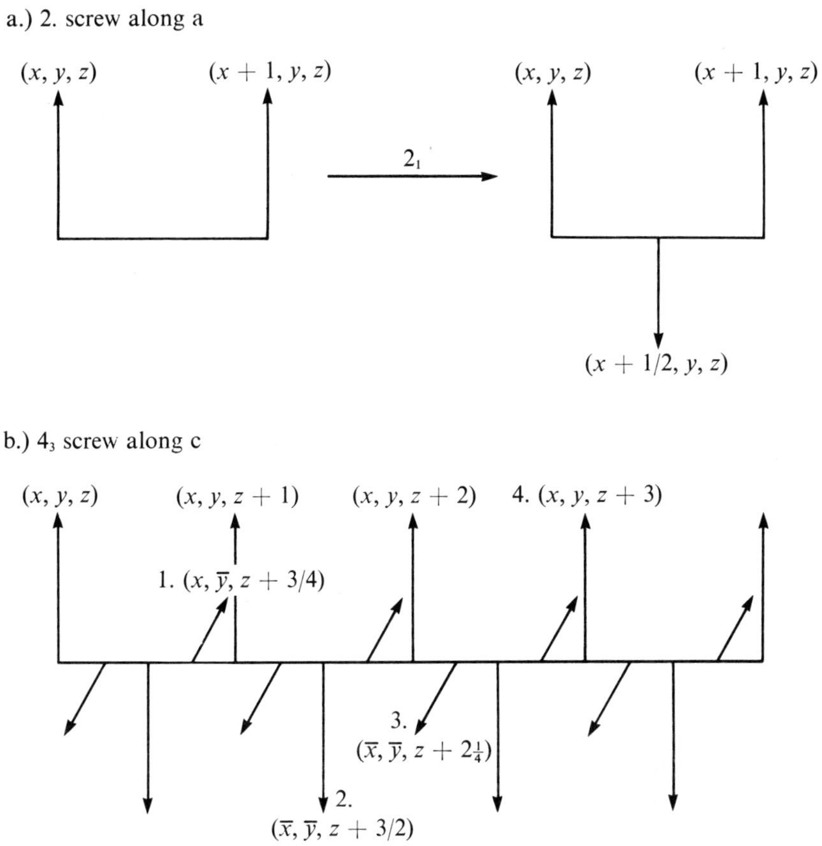

Figure 8.6
Effects of (a) 2_1 Screw along a, and (b) 4_3 Screw along c

8-8 THREE DIMENSIONAL SPACE GROUPS

When the new translational symmetry operations are combined with the symmetry operations that we have considered earlier, the 230 space groups result. These 230 space groups account for all the possible ways that identical points may be arranged in infinite lattices.

Let us first consider the combinations of the 32 point groups with the 14 Bravais lattices. To designate the space groups the letters *P*, *A*, *B*, *C*, *I*, and

F will be used in conjunction with the point group symbols. This practice led to the designations in Figure 8.5.

There are two triclinic point groups, and since triclinic unit cells are always chosen to be primitive, there are two triclinic space groups, those being $P1$ and $P\bar{1}$. For monoclinic systems we have three point groups, 2, *m*, and 2/*m*, and the possibility of both primitive cells and *c*-centered cells. Thus six monoclinic space groups are $P2$, Pm, $P2/m$, $C2$, Cm, and $C2/m$. Proceeding in this manner and taking care to include all possible combinations without duplications, 72 space groups are generated. These are listed in Table 8.4 along with the space groups that are generated by systematically replacing the mirror planes with glide planes and the rotation

Table 8.4
The 230 Space Groups

Point group	*Space group*	*Point group*	*Space group*	*Point group*	*Space group*
Triclinic		Orthorhombic		Orthorhombic	
1	$P1$		$I2_12_12_1$		$Pccm$
1	$P\bar{1}$	$mm2$	$Pmm2$		$Pban$
Monoclinic			$Pmc2_1$		$Pmma$
2	$P2$		$Pcc2$		$Pnna$
	$P2_1$		$Pma2$		$Pmna$
	$C2$		$Pca2_1$		$Pcca$
m	Pm		$Pnc2$		$Pbam$
	Pc		$Pmn2_1$		$Pccn$
	Cm		$Pba2$		$Pbcm$
	Cc		$Pna2_1$		$Pnnm$
$2/m$	$P2/m$		$Pnn2$		$Pmmn$
	$P2_1/m$		$Cmm2$		$Pbcn$
	$C2/m$		$Cmc2_1$		$Pbca$
	$P2/c$		$Ccc2$		$Pnma$
	$P2_1/c$		$Amm2$		$Cmcm$
	$C2/c$		$Abm2$		$Cmca$
Orthorhombic			$Ama2$		$Cmmm$
222	$P222$		$Aba2$		$Cccm$
	$P222_1$		$Fmm2$		$Cmma$
	$P2_12_12$		$Fdd2$		$Ccca$
	$P2_12_12_1$		$Imm2$		$Fmmm$
	$C222_1$		$Iba2$		$Fddd$
	$C222$		$Ima2$		$Immm$
	$F222$	mmm	$Pmmm$		$Ibam$
	$I222$		$Pnnn$		$Ibca$

Table 8.4 *(continued)*

Point group	*Space group*	*Point group*	*Space group*	*Point group*	*Space group*
Orthorhombic		Tetragonal		Trigonal	
	$Imma$		$I4_1cd$		$R3$
Tetragonal		$\bar{4}2m$	$P\bar{4}2m$	$\bar{3}$	$P\bar{3}$
4	$P4$		$P\bar{4}2c$		$R\bar{3}$
	$P4_1$		$P\bar{4}2_1m$	32	$P312$
	$P4_2$		$P\bar{4}2_1c$		$P321$
	$P4_3$		$P\bar{4}m2$		$P3_112$
	$I4$		$P\bar{4}c2$		$P3_121$
	$I4_1$		$P\bar{4}b2$		$P3_212$
$\bar{4}$	$P\bar{4}$		$P\bar{4}n2$		$P3_221$
	$I\bar{4}$		$I\bar{4}m2$		$R32$
$4/m$	$P4/m$		$I\bar{4}c2$	$3m$	$P3m1$
	$P4_2/m$		$I\bar{4}2m$		$P31m$
	$P4/n$		$I\bar{4}2d$		$P3c1$
	$P4_2/n$	$4/mmm$	$P4/mmm$		$P31c$
	$I4/m$		$P4/mcc$		$R3m$
	$I4_1/a$		$P4/nbm$		$R3c$
422	$P422$		$P4/nnc$	$\bar{3}m$	$P\bar{3}1m$
	$P42_12$		$P4/mbm$		$P\bar{3}1c$
	$P4_122$		$P4/mnc$		$P\bar{3}m1$
	$P4_12_12$		$P4/nmm$		$P\bar{3}c1$
	$P4_222$		$P4/ncc$		$R\bar{3}m$
	$P4_22_12$		$P4_2/mmc$		$R\bar{3}c$
	$P4_322$		$P4_2/mcm$	Hexagonal	
	$P4_32_12$		$P4_2/nbc$	6	$P6$
	$I422$		$P4_2/nnm$		$P6_1$
	$I4_122$		$P4_2/mbc$		$P6_5$
$4mm$	$P4mm$		$P4_2/mnm$		$P6_2$
	$P4bm$		$P4_2/nmc$		$P6_4$
	$P4_2cm$		$P4_2/ncm$		$P6_3$
	$P4_2nm$		$I4/mmm$	$\bar{6}$	$P\bar{6}$
	$P4cc$		$I4/mcm$	$6/m$	$P6/m$
	$P4nc$		$I4_1/amd$		$P6_3/m$
	$P4_2mc$		$I4_1/acd$	622	$P622$
	$P4_2bc$	Trigonal			$P6_122$
	$I4mm$	3	$P3$		$P6_522$
	$I4cm$		$P3_1$		$P6_222$
	$I4_1md$		$P3_2$		$P6_422$

Table 8.4 *(continued)*

Point group	*Space group*	*Point group*	*Space group*	*Point group*	*Space group*
Hexagonal		Cubic		Cubic	
	$P6_322$		$P2_13$	$\bar{4}3m$	$P\bar{4}3m$
$6mm$	$P6mm$		$I2_13$		$F\bar{4}3m$
	$P6cc$	$m3$	$Pm3$		$I\bar{4}3m$
	$P6_3cm$		$Pn3$		$P\bar{4}3n$
	$P6_3mc$		$Fm3$		$F\bar{4}3c$
$\bar{6}m2$	$P\bar{6}m2$		$Fd3$		$I\bar{4}3d$
	$P\bar{6}c2$		$Im3$	$m3m$	$Pm3m$
	$P\bar{6}2m$		$Pa3$		$Pn3n$
	$P\bar{6}2c$		$Ia3$		$Pm3n$
$6/mmm$	$P6/mmm$	432	$P432$		$Pn3m$
	$P6/mcc$		$P4_232$		$Fm3m$
	$P6_3/mcm$		$F432$		$Fm3c$
	$P6_3/mmc$		$F4_132$		$Fd3m$
Cubic			$I432$		$Fd3c$
23	$P23$		$P4_332$		$Im3m$
	$F23$		$P4_132$		$Ia3d$
	$I23$		$I4_132$		

axes with screw axes. In this latter regard, recall that each position of the component in the three-place symbol for the point group refers to a different direction. Thus, the symbol *Pbca* is for an orthorhombic space group. The subscript *b* in the first place designates a *b*-glide perpendicular to *a*, the letter *c* in the second place designates a *c*-glide perpendicular to *b*, and the letter *a* in the third place represents an *a*-glide perpendicular to *c*. Reference to Table 8.4 will show that 8 space groups are generated from the point group *mmm*.

Let us now see how many space groups can be derived from the point group $2/m$. First, we can have both *P* and *C* unit cells. Also, the two-fold rotation axis can be replaced by a two-fold screw axis and the mirror plane by a *c*-glide perpendicular to *b*. The eight permutations of the symmetry operations are

$$P2/m \qquad P2/c$$
$$C2/m \qquad C2/c$$
$$P2_1/m \qquad P2_1/c$$
$$C2_1/m \qquad C2_1/c$$

There are two duplications here because a two-fold rotation for a c-centered monoclinic lattice gives a result that is equivalent to the 2_1 screw axis parallel to b. The six unique space groups are $P2/m$, $C2/m$, $P2_1/m$, $P2_1/c$, $P\,2/c$, and $C2/c$.

8-9 INTERNATIONAL TABLES FOR X-RAY CRYSTALLOGRAPHY

Complete information about the 230 space groups has been collected in the International Tables for X-Ray Crystallography. These tables are indispensable for those interested in the symmetry properties of the crystalline state. For our purposes, the tables of the equivalent positions within the space groups are of primary importance. In Table 8.5 the list of general

Table 8.5

Equivalent Positions in the Space Group $P4_2/ncm$

Number of positions, Wyckoff notation, and point symmetry			*Co-ordinates of equivalent positions*
16	j	1	$x, y, z;\ \bar{x}, y, \tfrac{1}{2}+z;\ \tfrac{1}{2}-x, \tfrac{1}{2}+y, \bar{z};\ \tfrac{1}{2}+x, \tfrac{1}{2}+y, \tfrac{1}{2}-z;$
			$\bar{x}, \bar{y}, z;\ x, \bar{y}, \tfrac{1}{2}+z;\ \tfrac{1}{2}+x, \tfrac{1}{2}-y, \bar{z};\ \tfrac{1}{2}-x, \tfrac{1}{2}-y, \tfrac{1}{2}-z;$
			$\bar{y}, x, \bar{z};\ y, x, \tfrac{1}{2}-z;\ \tfrac{1}{2}+y, \tfrac{1}{2}+x, z;\ \tfrac{1}{2}-y, \tfrac{1}{2}+x, \tfrac{1}{2}+z;$
			$y, \bar{x}, \bar{z};\ \bar{y}, \bar{x}, \tfrac{1}{2}-z;\ \tfrac{1}{2}-y, \tfrac{1}{2}-x, z;\ \tfrac{1}{2}+y, \tfrac{1}{2}-x, \tfrac{1}{2}+z.$
8	i	m	$x, \tfrac{1}{2}+x, z;\ \tfrac{1}{2}+x, x, \tfrac{1}{2}-z;\ x, \tfrac{1}{2}-x, \tfrac{1}{2}+z;\ \tfrac{1}{2}+x, \bar{x}, \bar{z};$
			$\bar{x}, \tfrac{1}{2}-x, z;\ \tfrac{1}{2}-x, \bar{x}, \tfrac{1}{2}-z;\ \bar{x}, \tfrac{1}{2}+x, \tfrac{1}{2}+z;\ \tfrac{1}{2}-x, x, \bar{z}.$
8	h	2	$x, x, \tfrac{3}{4};\ \bar{x}, \bar{x}, \tfrac{3}{4};\ \tfrac{1}{2}+x, \tfrac{1}{2}+x, \tfrac{3}{4};\ \tfrac{1}{2}-x, \tfrac{1}{2}-x, \tfrac{3}{4};$
			$x, \bar{x}, \tfrac{1}{4};\ \bar{x}, x, \tfrac{1}{4};\ \tfrac{1}{2}+x, \tfrac{1}{2}-x, \tfrac{1}{4};\ \tfrac{1}{2}-x, \tfrac{1}{2}+x, \tfrac{1}{4}.$
8	g	2	$x, x, \tfrac{1}{4};\ \bar{x}, \bar{x}, \tfrac{1}{4};\ \tfrac{1}{2}+x, \tfrac{1}{2}+x, \tfrac{1}{4};\ \tfrac{1}{2}-x, \tfrac{1}{2}-x, \tfrac{1}{4};$
			$x, \bar{x}, \tfrac{3}{4};\ \bar{x}, x, \tfrac{3}{4};\ \tfrac{1}{2}+x, \tfrac{1}{2}-x, \tfrac{3}{4};\ \tfrac{1}{2}-x, \tfrac{1}{2}+x, \tfrac{3}{4}.$
8	f	2	$0, 0, z;\ 0, 0, \bar{z};\ 0, 0, \tfrac{1}{2}+z;\ 0, 0, \tfrac{1}{2}-z;$
			$\tfrac{1}{2}, \tfrac{1}{2}, z;\ \tfrac{1}{2}, \tfrac{1}{2}, \bar{z};\ \tfrac{1}{2}, \tfrac{1}{2}, \tfrac{1}{2}+z;\ \tfrac{1}{2}, \tfrac{1}{2}, \tfrac{1}{2}-z.$
4	e	mm	$0, \tfrac{1}{2}, z;\ 0, \tfrac{1}{2}, \tfrac{1}{2}+z;\ \tfrac{1}{2}, 0, \bar{z};\ \tfrac{1}{2}, 0, \tfrac{1}{2}-z.$
4	d	$2/m$	$\tfrac{1}{4}, \tfrac{1}{4}, \tfrac{3}{4};\ \tfrac{3}{4}, \tfrac{3}{4}, \tfrac{3}{4};\ \tfrac{1}{4}, \tfrac{3}{4}, \tfrac{1}{4};\ \tfrac{3}{4}, \tfrac{1}{4}, \tfrac{1}{4}.$
4	c	$2/m$	$\tfrac{1}{4}, \tfrac{1}{4}, \tfrac{1}{4};\ \tfrac{3}{4}, \tfrac{3}{4}, \tfrac{1}{4};\ \tfrac{1}{4}, \tfrac{3}{4}, \tfrac{3}{4};\ \tfrac{3}{4}, \tfrac{1}{4}, \tfrac{3}{4}.$
4	b	$\bar{4}$	$0, 0, 0;\ 0, 0, \tfrac{1}{2};\ \tfrac{1}{2}, \tfrac{1}{2}, 0;\ \tfrac{1}{2}, \tfrac{1}{2}, \tfrac{1}{2}.$
4	a	222	$0, 0, \tfrac{1}{4};\ 0, 0, \tfrac{3}{4};\ \tfrac{1}{2}, \tfrac{1}{2}, \tfrac{1}{4};\ \tfrac{1}{2}, \tfrac{1}{2}, \tfrac{3}{4}.$

and special positions for the point group $P4_2/ncm$ is given. In the first column of the table we see the number of equivalent positions that have the coordinates listed in the last column. The Wyckoff notation for the site is given in the second column, and the point symmetry of the site is given in the third column. In the Wyckoff notation the highest symmetry site is designated by the letter *a*, with lower symmetry sites being assigned consecutive letters of the alphabet in order of decreasing symmetry. Thus, in the space group $P4_2/ncm$ there are four sites with symmetry 222, (D_2) these being the highest order symmetry sites in the group. The next highest order sites have $\bar{4}$ symmetry, and there are four such sites.

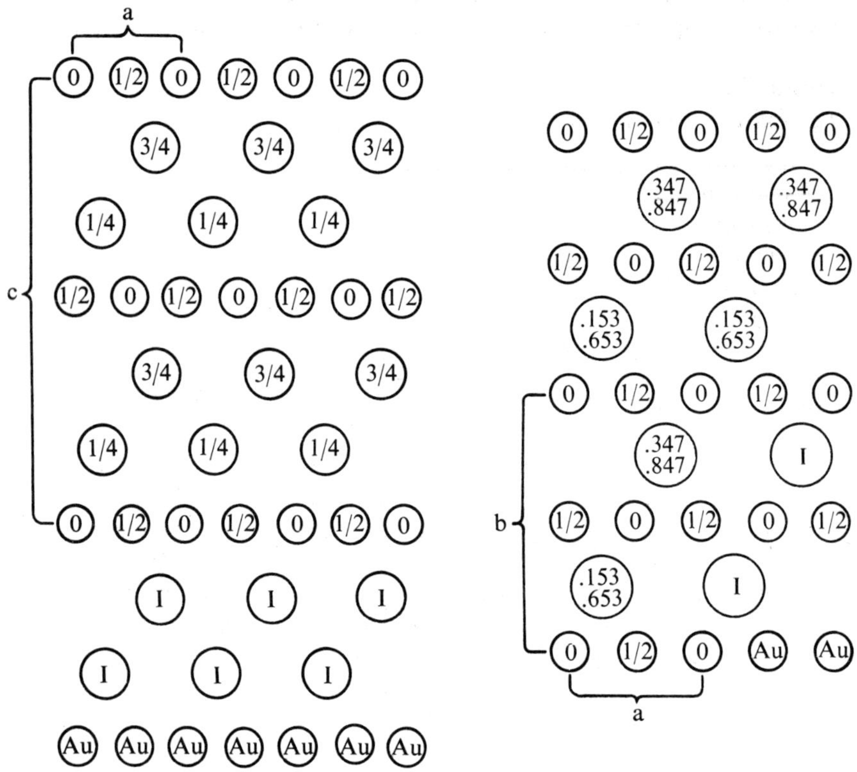

Figure 8.7
(a) View of the AuI Structure as Projected on the ac Plane, (b) as Projected on the ab Plane

As an example we shall describe the structure of AuI, which crystallizes in the space group $P4_2/ncm$ with four molecules per unit cell. The atoms are arranged in space as shown in Figure 8.7. It is instructive to examine the projections of this structure on the *ab* and the *ac* planes and to pick out the various symmetry elements. The 4_2 screw axis is readily identified parallel to the *c*-axis. The *n*-glide perpendicular to the 4_2 axis is more difficult to see in the projection on the *ac* plane, but becomes apparent when the projection on the *ab* plane is examined (Figure 8.7b). The view in Figure 8.7a reveals the *c*-glide perpendicular to *a* and an equivalent element perpendicular to *b*. Finally, the mirror plane perpendicular to the *ab* diagonal is found in Figure 8.7b.

Exercise 8-4

Calculate the distances from a given gold ion in AuI to each of its nearest neighbors, and from a given iodide ion to each of its nearest neighbors. Describe the coordination geometry of each ion. Obtain the necessary data from the International Tables and Wyckoff's text.*

8-10 CRYSTAL STRUCTURE OF POTASSIUM CHROMATE

Later in this chapter we will need to know the structural details for K_2CrO_4. From volume 3 of Wyckoff's series, *Crystal Structures,** we find that K_2CrO_4 crystallizes in the orthorhombic space group *Pnma* with four formula units in a unit cell having dimensions $a = 7.61$ Å, $b = 5.92$ Å, and $c = 10.40$ Å. The potassium and chromium ions occupy Wyckoff 4*c* positions, as do half of the oxygen atoms. The other half of the oxygen atoms are in 8*d* sites. From the International Tables we find that the 4*c* sites have symmetry *m* (C_s) while the 8*d* sites have symmetry 1 (C_1). If necessary, views of the crystal structure could be constructed for additional insight into the structural details; however, we do not require any additional information for the vibrational spectroscopy problem that we will consider.

Exercise 8-5

From the International Tables and Wyckoff's book find the coordinates of the ions in the unit cell of K_2CrO_4. Calculate the distances from each

*R. W. G. Wyckoff, *Crystal Structures*, 2d ed. (New York: Interscience Publishers, 1963).

chromium ion to all of its nearest neighbors. What is the coordination geometry of the chromium ion?

8-11 VIBRATIONAL SPECTRA OF SOLID SAMPLES

In Section 5-7 we pointed out that the selection rules for the activity of vibrational motions of solids in infrared and Raman spectra must be discussed in terms of the symmetry properties of the crystalline state. The two usual approaches to this problem are site symmetry analysis and factor group analysis. The procedure to be selected depends on the magnitude of the interactions between the molecules in the solid state. If the interactions are weak, then the site symmetry approach will probably be adequate. However, if the intermolecular interactions are strong, then the factor group analysis must be applied. The number of low energy vibrational bands present in a spectrum is an experimental guide to the appropriate technique. If there are more bands in the spectrum than would be expected on the basis of the site symmetry approach, then the factor group analysis must be used.

In the absence of experimental data it is difficult to predict which technique will be required. However, one can expect that molecules which are closely packed in the solid state and which have highly dipolar vibrational motions will interact strongly and will require the factor group analysis. Since dipole-dipole interactions vary as $1/r^3$, we may expect little interaction between molecules which are isolated from one another, and we may expect that the site symmetry analysis will be adequate in those instances.

The bands that are observed in the vibrational spectra of solids may be divided into two classes. One class includes those vibrational modes which arise from the motions of the atoms relative to one another in each of the molecules or polyatomic ions which constitute the lattice. These internal modes have been treated in detail in Chapter 5 and usually occur at energies rather close to those observed for the free molecule or ion. The additional modes which form the second class occur only for solids and are called lattice or external modes. Lattice modes are further subdivided into rotational and translational modes, and these arise from the rotations and translations of molecules or ions relative to one another. The fundamental lattice vibrations are found in the very low energy region of the spectrum. We shall now discuss the site symmetry analysis and factor group analysis and then apply the technique to some representative examples.

8-12 SITE SYMMETRY SELECTION RULES

The assumption that intermolecular coupling of the vibrational motions is negligible forms the basis for the site symmetry analysis. Therefore, the determination of the internal modes is identical to the procedure outlined in Section 5-5, but here the site symmetry of the molecule or polyatomic ion is used for the construction of the symmetry species. Since the site symmetries occupied by molecules or ions are frequently lower than the symmetry of the free molecule, the effect of the lower symmetry environment is to split some of the degeneracies, if any, of the normal modes and to relax some of the selection rules.

The site symmetry of a molecule or ion may be determined by reference to the site symmetry of the central atom. For example the site symmetry of the carbon atom of the carbonate ion in a given lattice will be the symmetry of the carbonate ion. The site symmetry of the central atom is available from the X-ray crystal structure data where most frequently the coordinates and Wyckoff positions are recorded.

It was pointed out in Section 5-7 that the two crystallographically different forms of $CaCO_3$ serve to illustrate the effect of different site symmetries on infrared spectra. The two crystallographic forms are calcite, in which the carbonate ion sits on a site with 32 (D_3) symmetry, and aragonite, in which the carbonate ion sits on a site with m symmetry. The free carbonate ion should have D_{3h} symmetry, and an analysis of the internal modes of the free ions reveals that the fundamental modes transform as $A_1' + A_2'' + 2E'$. In Table 8.6 we show the correlation between the irreducible representations in D_{3h} and the lower symmetry point groups D_3 and C_s. Reference to the character table for D_3 shows that the A_2 and E modes are infrared active, and the spectrum of calcite should be similar to the spectrum of the free carbonate ion as obtained in solution. However, in aragonite the symmetric stretching vibration, which is forbidden in D_{3h}, should be allowed since it correlates with A' in C_s and both the x and y components

Table 8.6
Correlation Table for D_{3h}, C_s, D_3

D_3	D_{3h}	C_s
A_1	A_1'	A'
A_2	A_2''	A''
E	E'	

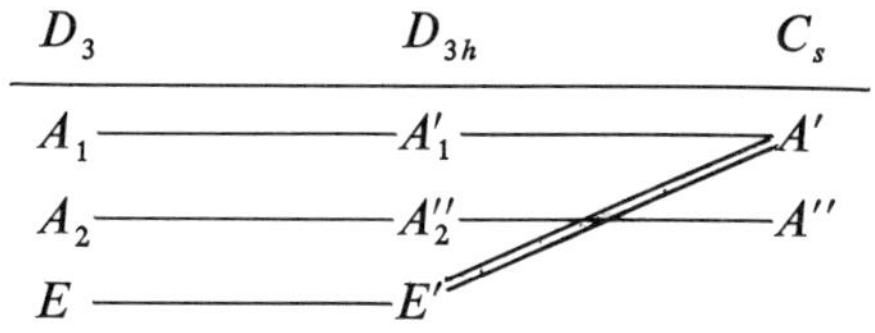

of the dipole moment operator transform as A' in this point group. In addition there are no doubly degenerate representations in C_s, and the degeneracies of the E' modes will be removed. In principle then, there could be five bands in the infrared spectrum of aragonite arising from internal fundamental modes. Only four bands are observed, presumably due to a very small splitting of one of the E' modes or due to a low intensity for one of the components of a split E' mode.

The symmetries of the translational and rotational motions of the molecules in the crystalline state are determined in the same manner discussed in Section 5-7. Since the translations transform as x, y, z, all three translational motions will always be infrared active. The infrared and Raman activity of the translational and rotational motions are determined in the usual manner, that is, if R_q ($q = x, y, z$) transforms as x, y, or z, the rotational motion is infrared active, and if R_q and/or T_q transform as a square or binary product of x, y, or z, the external mode is Raman active.

8-13 FACTOR GROUP ANALYSIS

If intermolecular dipolar coupling is very strong then factor group analysis must be used in order to explain vibrational spectra. Under the assumption of complete vibrational coupling, the vibrational modes are constructed from the motions of all the atoms in the unit cell, and not just from the motions of the atoms in the molecules or polyatomic ions which may be contained in the cell. Thus, we may expect that the technique used for the determination of the symmetry properties of the vibrational modes of the unit cell would be very similar to that used for isolated molecules, except that all of the atomic positions in the unit cell are used for the formation of the representation of the appropriate group. The group which describes the symmetry properties of a single unit cell is a factor group of the space group to which the crystal belongs.

The factor group is isomorphous with one of the 32 crystallographic point groups and may be obtained by deleting the superscript from the Schoenflies symbol for the space group. For example, the Schoenflies symbol for the space group $P4_2/nmc$ is D_{4h}^{15}. For this space group the factor group is D_{4h}, and the character table for D_{4h} is the appropriate character table for the factor group. However, it should be pointed out that although the point group and factor group character tables are identical, the operations of rotation and reflection in the point group may correspond to the operations of screw and glide in the factor group.

For an application of the factor group analysis we will use the correlation chart method. We will need from X-ray data the space group, Wyckoff positions (site symmetries) of the atoms in the unit cell, and the number of molecules in the unit cell. The correlation chart method utilizes the following steps:

1. The symmetries of the fundamental internal modes of vibration are determined for each of the polyatomic ions and molecules using the appropriate point group for the free ion or molecule.
2. The symmetries of the translational and rotational modes of the molecules and ions are determined.
3. The irreducible representations of the free ion point group are correlated with the irreducible representations of the site group.
4. The irreducible representations spanned by the translational modes of the monoatomic ions are determined using the site group, and these are correlated with the irreducible representations of the factor group.
6. The results of steps 4 and 5 are combined and three degrees of translational freedom, representing the translation of the entire unit cell, are subtracted, thus leaving the lattice or external vibrations. These three degrees of freedom are called acoustical modes, and they are not infrared active. The translational degrees of freedom to be subtracted are those spanned by x, y, z in the factor group.
7. Infrared and Raman spectral activity is determined in the usual manner using the character table for the point group isomorphous with the factor group.

We will now apply the seven steps of the factor group-correlation chart method to the analysis of the vibrational spectrum fo K_2CrO_4. As noted in Section 8.10 potassium chromate crystallizes in an orthorhombic system belonging to space group D_{2h}^{16}-*Pnma* with four formula units per unit cell. The potassium ions and the chromate ions sit on positions with $m\,(C_s)$ symmetry.

Step 1. The symmetry species for the fundamental modes of the chromate ion are determined. Here we shall assume T_d symmetry for the free ion. By the procedures described in Section 5-5 we find that

$$\Gamma_{\text{b.v.}} = A_1 + E + T_1 + 3T_2$$

From the character table it may be seen that

$$\Gamma_{\text{rot}} = T_1$$

and

$$\Gamma_{\text{trans}} = T_2$$

Thus

$$\Gamma_{\text{vib}} = A_1 + E + 2T_2$$

Since there are four chromate ions in the unit cell, there will be a total of 16 symmetry species and 36 fundamental internal modes.

Step 2. In step 1 we found that $\Gamma_{\text{rot}} = T_1$ and $\Gamma_{\text{trans}} = T_2$, and for the four ions we have four rotational symmetry species and four translational species, giving rise to the 24 external modes.

Step 3. The internal and external irreducible representations of the free ion are correlated with the proper irreducible representations of the site group. The Selected Correlation Tables in Appendix II are valuable and save time, although the correlations can be determined with little effort. The correlations are

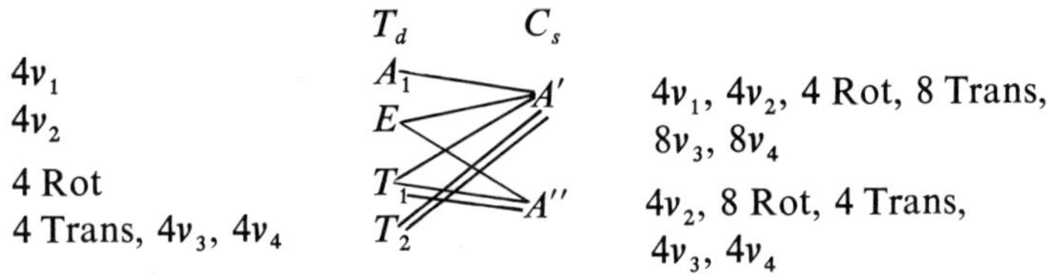

Step 4. The irreducible representation of the site group is now correlated with the irreducible representations of the factor group:

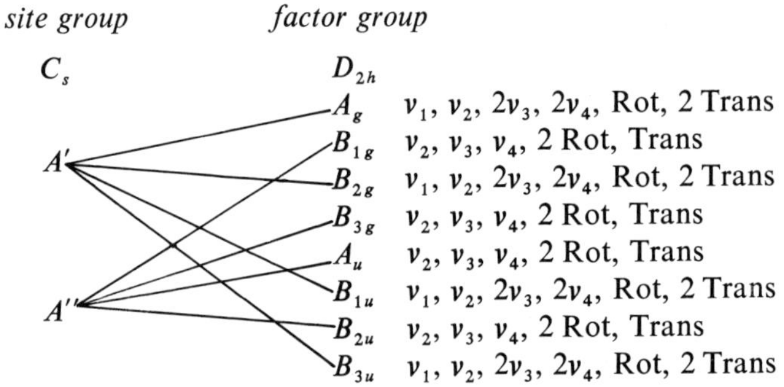

Step 5. The potassium ions sit on sites with m symmetry. In C_s x and y transform as A' and z transforms as A''. Thus, the eight potassium ions will give rise to $16A'$ and $8A''$ translational modes.

These are correlated in the following manner:

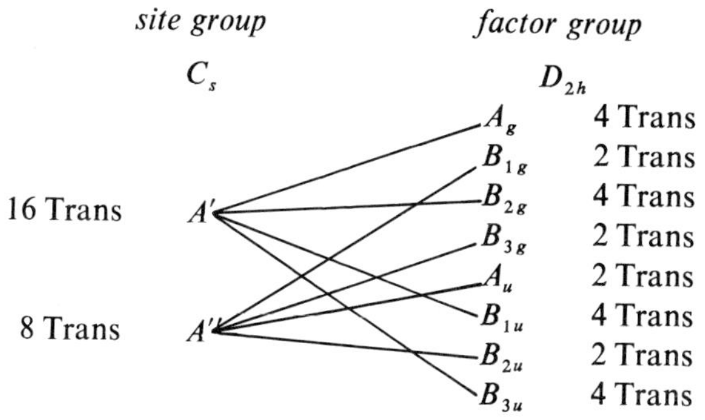

Step 6. Here we combine the results of steps 4 and 5 and subtract one translational mode each from B_{1u}, B_{2u}, and B_{3u}, because in D_{2h} x, y, and z transform as B_{1u}, B_{2u}, and B_{3u}. These modes are the acoustical modes and have no spectral activity.

Step 7. The spectral activity is determined by referring to the character table for the point group isomorphous with the factor group, in this case D_{2h}. We find that the fundamental vibrations that transform as the irreducible representations B_{1u}, B_{2u}, and B_{3u} are infrared active only, and that those belonging to A_g, B_{1g}, B_{2g}, and B_{3g} are Raman active. These data have been indicated in Table 8.7 with a √ for an allowed mode and an x for a forbidden mode.

Table 8.7
Selection Rules

symmetry species	*motions*	*infrared*	*Raman*
A_g	ν_1, ν_2, $2\nu_3$, $2\nu_4$, Rot, 6 Trans	x	√
B_{1g}	ν_2, ν_3, ν_4, 2 Rot, 3 Trans	x	√
B_{2g}	ν_1, ν_2, $2\nu_3$, $2\nu_4$, Rot, 6 Trans	x	√
B_{3g}	ν_2, ν_3, ν_4, 2 Rot, 3 Trans	x	√
A_u	ν_1, ν_3, ν_4, 2 Rot, 3 Trans	x	x
B_{1u}	ν_1, ν_2, $2\nu_3$, $2\nu_4$, Rot, 5 Trans	√	x
B_{2u}	ν_2, ν_3, ν_4, 2 Rot, 2 Trans	√	x
B_{3u}	ν_1, ν_2, $2\nu_3$, $2\nu_4$, Rot, 5 Trans	√	x

It is important to note that there will be no coincidences in the infrared and Raman spectra of K_2CrO_4. This mutual exclusion arises because the crystal is centrosymmetric. The free chromate ion does not have a center of symmetry, and the fundamental modes which transform as T_2 are both Raman and infrared active. The numbering system for the fundamental internal modes used for the factor group is a matter of convenience and permits us to keep track of the internal modes, but does not imply that these modes arise only from independent molecules or polyatomic ions. It may be recalled that the initial assumption was that intermolecular coupling dominates intramolecular coupling.

From a study of a single crystal sample for the Raman spectrum and a powdered sample for the infrared spectrum, the internal modes and many of the lattice modes of K_2CrO_4 have been tentatively identified. The results are summarized in Table 8.8. The splitting in the A_1 mode (T_d) due to intermolecular coupling was observed in the infrared spectrum only at 4.2°K, but at room temperature only one band was observed. It can be seen that

Table 8.8 (A)

Vibrational Spectrum of K_2CrO_4*; Internal Modes*

Raman (Aqueous solution)	*Site Symmetry*	*Raman (crystal)*	*Infrared (powder)*
T_d	C_s		
		919 B_{2g}	920 $B_{1u} + B_{3u}$
	A'	905 A_g	
884 T_2		884 B_{2g}	890 $B_{1u} + B_{3u}$
		868 A_g	
	A''	881 B_{1g}	— A_u (forbidden)
		877 B_{3g}	876 B_{2u}
			856
847 A_1	A'	852 $A_g + B_{2g}$	845 $B_{1u} + B_{3u}$
		398 B_{2g}	398 $B_{1u} + B_{3u}$
	A'	397 A_g	
368 T_2		388 B_{2g}	385 B_{2u}
		387 A_g	
	A''	393 B_{3g}	— A_u (forbidden)
		388 B_{1g}	385 B_{2u}
	A'	352 B_{2g}	
348 E		347 A_g	
	A''	351 B_{3g}	
		348 B_{1g}	343 $B_{2u} + (B_{1u} + B_{3u})$

Table 8.8 (B)

Vibrational Spectrum of K_2CrO_4*; Lattice Modes*

Raman powder	single crystal			
	A_{1g}	B_{1g}	B_{2g}	B_{3g}
60*w*		58*wm*		
72*vvw*				
84*w*	84*w*	87*vw*		
95*vw*		94*wm*		94*vvw*
125*vvw*			120*vw*	119*w*
144*vvw*		138*vvw*		
168*wm*	164*wm*			164*vvw*
184*vvw*	181*w*			182*vw*
Infrared	B_{1u}	B_{2u}	B_{3u}	
	58*vw*			
		67*wm**		
	102*ms*		104*wm**	
		113*ms*		
	141*s*	139*s*	142*sh*	
	163*s*		161*s*	
	180*sh*		180*sh*	

*Observed in the reflection spectrum. All other bands were from the absorption spectrum.

for all the other T_d modes found in the Raman spectrum the splitting is predicted by the factor group method. Of the 24 Raman lines predicted for the lattice modes by the analysis, only 12 were observed, with the deficiency being mainly A_g and B_{2g} species. Twelve of the 16 infrared absorptions were observed. However, there is little agreement between the data and assignments given by Carter*, and Adams, *et al.*†

8-14 VIBRATIONAL SPECTRUM OF π-BENZENETRICARBONYLCHROMIUM

Free π-benzenetricarbonylchromium would be expected to have C_{3v} symmetry and, therefore, should exhibit two bands in the infrared spectrum.

*R. L. Carter, *J. Chem. Ed.*, *48*, 297 (1971).
†D. M. Adams, M. A. Hooper, and M. H. Lloyd, *J. Chem. Soc. A*, 946 (1971).

In fact, a carbon tetrachloride solution of the compound does exhibit two bands at 1909 and 1978 cm^{-1}, respectively. In the solid state the following absorption bands are observed:

Infrared	1966			1879		1858
Raman		1945	1887		1865	

The appearance of six bands and the absence of coincidences between the infrared and Raman spectra suggest that intermolecular vibrational coupling is very strong, and we expect that the compound crystallizes in a centrosymmetric space group.

From the X-ray data it is known that π-benzenetricarbonylchromium crystallizes in the monoclinic space group C_{2h}^2-$P2_1/m$ with two molecules per unit cell. Using the correlation chart method for the factor group analysis, the correlation chart in Table 8.9 results, where it is seen that the selection rules predict that three infrared bands and three Raman bands will arise from the carbonyl stretching vibrations with no coincidences, in agreement with the experimental observations.

Table 8.9

Correlation Chart for π-$C_6H_6Cr(CO)_3$

free molecule, C_{3v}	*site group,* C_s	*factor group,* C_{2h}	*Spectral Activity*
$2v_1$ A_1	A'	A_g v_1, v_2	Raman
		B_g v_2	Raman
$2v_2$ E	A''	A_u v_2	Infrared
		B_u v_1, v_2	Infrared

Unfortunately, the problem does not always lend itself to such a ready solution. Among the difficulties are accidental coincidences in the infrared and Raman spectra which lead to fewer bands than would be predicted.

Exercise 8-6

Hexamethylbenzenetricarbonylchromium crystallizes in the orthorhombic space group D_{2h}^{15}-$Pbca$ with eight molecules per unit cell. What is the site symmetry of the chromium atom (see the X-ray structure paper and the International Tables)? Predict the number of allowed carbonyl stretching vibrations in the infrared and Raman spectra. Should coincidences be observed? Provide an explanation for the experimental observation of five infrared and five non-coincidental Raman bands in the carbonyl region.

Bibliography

There are a number of authoritative books and reference works dealing with the symmetry and structure of the solid state. Useful books on crystallography include:

D. E. Sands, *Introduction to Crystallography*, New York: W. A. Benjamin, 1969.

G. H. Stout and L. H. Jensen, *X-Ray Structure Determination*, New York: Macmillan, 1968.

International Tables for X-Ray Crystallography. Birmingham, England: Kynoch Press, 1962.

Much structural data may be found in the following series:

R. W. G. Wyckoff, *Crystal Structures*, 2d ed. New York: Interscience, 1963.

Factor group analysis is discussed by

A. Bhagavantam and T. Venkatarayudu, *Theory of Groups and Its Application to Physical Problems*. New York: Academic Press, 1969.

Appendix I

CHARACTER TABLES

1. THE NONAXIAL GROUPS

C_1	E
A	1

C_s	E	σ_h		
A'	1	1	x, y, R_z	x^2, y^2, z^2, xy
A''	1	-1	z, R_x, R_y	yz, xz

C_i	E	i		
A_g	1	1	R_x, R_y, R_z	$x^2, y^2, z^2, xy, xz, yz$
A_u	1	-1	x, y, z	

2. THE C_n GROUPS

C_2	E	C_2		
A	1	1	z, R_z	x^2, y^2, z^2, xy
B	1	-1	x, y, R_x, R_y	yz, xz

2. THE C_n GROUPS *(continued)*

C_3	E	C_3	C_3^2		$\varepsilon = \exp(2\pi i/3)$
A	1	1	1	z, R_z	$x^2 + y^2, z^2$
E	$\begin{Bmatrix} 1 \\ 1 \end{Bmatrix}$	$\begin{matrix} \varepsilon \\ \varepsilon^* \end{matrix}$	$\begin{Bmatrix} \varepsilon^* \\ \varepsilon \end{Bmatrix}$	$(x, y)(R_x, R_y)$	$(x^2 - y^2, xy)(yz, xz)$

C_4	E	C_4	C_2	C_4^3		
A	1	1	1	1	z, R_z	$x^2 + y^2, z^2$
B	1	−1	1	−1		$x^2 - y^2, xy$
E	$\begin{Bmatrix} 1 \\ 1 \end{Bmatrix}$	$\begin{matrix} i \\ -i \end{matrix}$	$\begin{matrix} -1 \\ -1 \end{matrix}$	$\begin{Bmatrix} -i \\ i \end{Bmatrix}$	$(x, y)(R_x, R_y)$	(yz, xz)

C_5	E	C_5	C_5^2	C_5^3	C_5^4		$\varepsilon = \exp(2\pi i/5)$
A	1	1	1	1	1	z, R_z	$x^2 + y^2, z^2$
E_1	$\begin{Bmatrix} 1 \\ 1 \end{Bmatrix}$	$\begin{matrix} \varepsilon \\ \varepsilon^* \end{matrix}$	$\begin{matrix} \varepsilon^2 \\ \varepsilon^{2*} \end{matrix}$	$\begin{matrix} \varepsilon^{2*} \\ \varepsilon^2 \end{matrix}$	$\begin{Bmatrix} \varepsilon^* \\ \varepsilon \end{Bmatrix}$	$(x, y)(R_x, R_y)$	(yz, xz)
E_2	$\begin{Bmatrix} 1 \\ 1 \end{Bmatrix}$	$\begin{matrix} \varepsilon^2 \\ \varepsilon^{2*} \end{matrix}$	$\begin{matrix} \varepsilon^* \\ \varepsilon \end{matrix}$	$\begin{matrix} \varepsilon \\ \varepsilon^* \end{matrix}$	$\begin{Bmatrix} \varepsilon^{2*} \\ \varepsilon^2 \end{Bmatrix}$		$(x^2 - y^2, xy)$

C_6	E	C_6	C_3	C_2	C_3^2	C_6^5		$\varepsilon = \exp(2\pi i/6)$
A	1	1	1	1	1	1	z, R_z	$x^2 + y^2, z^2$
B	1	−1	1	−1	1	−1		
E_1	$\begin{Bmatrix} 1 \\ 1 \end{Bmatrix}$	$\begin{matrix} \varepsilon \\ \varepsilon^* \end{matrix}$	$\begin{matrix} -\varepsilon^* \\ -\varepsilon \end{matrix}$	$\begin{matrix} -1 \\ -1 \end{matrix}$	$\begin{matrix} -\varepsilon \\ -\varepsilon^* \end{matrix}$	$\begin{Bmatrix} \varepsilon^* \\ \varepsilon \end{Bmatrix}$	(x, y) (R_x, R_y)	(xz, yz)
E_2	$\begin{Bmatrix} 1 \\ 1 \end{Bmatrix}$	$\begin{matrix} -\varepsilon^* \\ -\varepsilon \end{matrix}$	$\begin{matrix} -\varepsilon \\ -\varepsilon^* \end{matrix}$	$\begin{matrix} 1 \\ 1 \end{matrix}$	$\begin{matrix} -\varepsilon^* \\ -\varepsilon \end{matrix}$	$\begin{Bmatrix} -\varepsilon \\ -\varepsilon^* \end{Bmatrix}$		$(x^2 - y^2, xy)$

2. THE C_n GROUPS *(continued)*

C_7	E	C_7	C_7^2	C_7^3	C_7^4	C_7^5	C_7^6		$\varepsilon = \exp(2\pi i/7)$
A	1	1	1	1	1	1	1	z, R_z	$x^2 + y^2, z^2$
E_1	1	ε	ε^2	ε^3	ε^{3*}	ε^{2*}	ε^*	$(x, y)(R_x, R_y)$	(xz, yz)
	1	ε^*	ε^{2*}	ε^{3*}	ε^3	ε^2	ε		
E_2	1	ε^2	ε^{3*}	ε^*	ε	ε^3	ε^{2*}		$(x^2 - y^2, xy)$
	1	ε^{2*}	ε^3	ε	ε^*	ε^{3*}	ε^2		
E_3	1	ε^3	ε^*	ε^2	ε^{2*}	ε	ε^{3*}		
	1	ε^{3*}	ε	ε^{2*}	ε^2	ε^*	ε^3		

C_8	E	C_8	C_4	C_2	C_4^3	C_8^3	C_8^5	C_8^7		$\varepsilon = \exp(2\pi i/8)$
A	1	1	1	1	1	1	1	1	z, R_z	$x^2 + y^2, z^2$
B	1	-1	1	1	1	-1	-1	-1		
E_1	1	ε	i	-1	$-i$	$-\varepsilon^*$	$-\varepsilon$	ε^*	$(x, y)(R_x, R_y)$	(xz, yz)
	1	ε^*	$-i$	-1	i	$-\varepsilon$	$-\varepsilon^*$	ε		
E_2	1	i	-1	1	-1	$-i$	i	$-i$		$(x^2 - y^2, xy)$
	1	$-i$	-1	1	-1	i	$-i$	i		
E_3	1	$-\varepsilon$	i	-1	$-i$	ε^*	ε	$-\varepsilon^*$		
	1	$-\varepsilon^*$	$-i$	-1	i	ε	ε^*	$-\varepsilon$		

3. THE D_n GROUPS

D_2	E	$C_2(z)$	$C_2(y)$	$C_2(x)$		
A	1	1	1	1		x^2, y^2, z^2
B_1	1	1	−1	−1	z, R_z	xy
B_2	1	−1	1	−1	y, R_y	xz
B_3	1	−1	−1	1	x, R_x	yz

D_3	E	$2C_3$	$3C_2$		
A_1	1	1	1		$x^2 + y^2, z^2$
A_2	1	1	−1	z, R_z	
E	2	−1	0	$(x, y)(R_x, R_y)$	$(x^2 - y^2, xy)(xz, yz)$

D_4	E	$2C_4$	$C_2\ (= C_4^2)$	$2C_2'$	$2C_2''$		
A_1	1	1	1	1	1		$x^2 + y^2, z^2$
A_2	1	1	1	−1	−1	z, R_z	
B_1	1	−1	1	1	−1		$x^2 - y^2$
B_2	1	−1	1	−1	1		xy
E	2	0	−2	0	0	$(x, y)(R_x, R_y)$	(xz, yz)

D_5	E	$2C_5$	$2C_5^2$	$5C_2$		
A_1	1	1	1	1		$x^2 + y^2, z^2$
A_2	1	1	1	−1	z, R_z	
E_1	2	2 cos 72°	2 cos 144°	0	$(x, y)(R_x, R_y)$	(xz, yz)
E_2	2	2 cos 144°	2 cos 72°	0		$(x^2 - y^2, xy)$

D_6	E	$2C_6$	$2C_3$	C_2	$3C_2'$	$3C_2''$		
A_1	1	1	1	1	1	1		$x^2 + y^2, z^2$
A_2	1	1	1	1	−1	−1	z, R_z	
B_1	1	−1	1	−1	1	−1		
B_2	1	−1	1	−1	−1	1		
E_1	2	1	−1	−2	0	0	$(x, y)(R_x, R_y)$	(xz, yz)
E_2	2	−1	−1	2	0	0		$(x^2 - y^2, xy)$

4. THE C_{nv} GROUPS

C_{2v}	E	C_2	$\sigma_v(xz)$	$\sigma_v'(yz)$		
A_1	1	1	1	1	z	x^2, y^2, z^2
A_2	1	1	−1	−1	R_z	xy
B_1	1	−1	1	−1	x, R_y	xz
B_2	1	−1	−1	1	y, R_x	yz

4. THE C_{nv} GROUPS *(continued)*

C_{3v}	E	$2C_3$	$3\sigma_v$		
A_1	1	1	1	z	$x^2 + y^2, z^2$
A_2	1	1	−1	R_z	
E	2	−1	0	$(x, y)(R_x, R_y)$	$(x^2 - y^2, xy)(xz, yz)$

C_{4v}	E	$2C_4$	C_2	$2\sigma_v$	$2\sigma_d$		
A_1	1	1	1	1	1	z	$x^2 + y^2, z^2$
A_2	1	1	1	−1	−1	R_z	
B_1	1	−1	1	1	−1		$x^2 - y^2$
B_2	1	−1	1	−1	1		xy
E	2	0	−2	0	0	$(x, y)(R_x, R_y)$	(xz, yz)

C_{5v}	E	$2C_5$	$2C_5^2$	$5\sigma_v$		
A_1	1	1	1	1	z	$x^2 + y^2, z^2$
A_2	1	1	1	−1	R_z	
E_1	2	2 cos 72°	2 cos 144°	0	$(x, y)(R_x, R_y)$	(xz, yz)
E_2	2	2 cos 144°	2 cos 72°	0		$(x^2 - y^2, xy)$

C_{6v}	E	$2C_6$	$2C_3$	C_2	$3\sigma_v$	$3\sigma_d$		
A_1	1	1	1	1	1	1	z	$x^2 + y^2, z^2$
A_2	1	1	1	1	−1	−1	R_z	
B_1	1	−1	1	−1	1	−1		
B_2	1	−1	1	−1	−1	1		
E_1	2	1	−1	−2	0	0	$(x, y)(R_x, R_y)$	(xz, yz)
E_2	2	−1	−1	2	0	0		$(x^2 - y^2, xy)$

5. THE C_{nh} GROUPS

C_{2h}	E	C_2	i	σ_h		
A_g	1	1	1	1	R_z	x^2, y^2, z^2, xy
B_g	1	−1	1	−1	R_x, R_y	xz, yz
A_u	1	1	−1	−1	z	
B_u	1	−1	−1	1	x, y	

C_{3h}	E	C_3	C_3^2	σ_h	S_3	S_3^5		$\varepsilon = \exp(2\pi i/3)$
A'	1	1	1	1	1	1	R_z	$x^2 + y^2, z^2$
E'	$\begin{cases} 1 \\ 1 \end{cases}$	$\begin{matrix} \varepsilon \\ \varepsilon^* \end{matrix}$	$\begin{matrix} \varepsilon^* \\ \varepsilon \end{matrix}$	$\begin{matrix} 1 \\ 1 \end{matrix}$	$\begin{matrix} \varepsilon \\ \varepsilon^* \end{matrix}$	$\begin{matrix} \varepsilon^* \\ \varepsilon \end{matrix}$	(x, y)	$(x^2 - y^2, xy)$
A''	1	1	1	−1	−1	−1	z	
E''	$\begin{cases} 1 \\ 1 \end{cases}$	$\begin{matrix} \varepsilon \\ \varepsilon^* \end{matrix}$	$\begin{matrix} \varepsilon^* \\ \varepsilon \end{matrix}$	$\begin{matrix} -1 \\ -1 \end{matrix}$	$\begin{matrix} -\varepsilon \\ -\varepsilon^* \end{matrix}$	$\begin{matrix} -\varepsilon^* \\ -\varepsilon \end{matrix}$	(R_x, R_y)	(xz, yz)

5. THE C_{nh} GROUPS (continued)

C_{4h}	E	C_4	C_2	C_4^3	i	S_4^3	σ_h	S_4		
A_g	1	1	1	1	1	1	1	1	R_z	$x^2 + y^2, z^2$
B_g	1	−1	1	−1	1	−1	1	−1		$x^2 - y^2, xy$
E_g	{1	i	−1	$-i$	1	i	−1	$-i$}	(R_x, R_y)	(xz, yz)
	{1	$-i$	−1	i	1	$-i$	−1	i}		
A_u	1	1	1	1	−1	−1	−1	−1	z	
B_u	1	−1	1	−1	−1	1	−1	1		
E_u	{1	i	−1	$-i$	−1	$-i$	1	i}	(x, y)	
	{1	$-i$	−1	i	−1	i	1	$-i$}		

C_{5h}	E	C_5	C_5^2	C_5^3	C_5^4	σ_h	S_5	S_5^7	S_5^3	S_5^9		$\varepsilon = \exp(2\pi i/5)$
A'	1	1	1	1	1	1	1	1	1	1	R_z	$x^2 + y^2, z^2$
E_1'	{1	ε	ε^2	ε^{2*}	ε^*	1	ε	ε^2	ε^{2*}	ε^*}	(x, y)	
	{1	ε^*	ε^{2*}	ε^2	ε	1	ε^*	ε^{2*}	ε^2	ε}		
E_2'	{1	ε^2	ε^*	ε	ε^{2*}	1	ε^2	ε^*	ε	ε^{2*}}		$(x^2 - y^2, xy)$
	{1	ε^{2*}	ε	ε^*	ε^2	1	ε^{2*}	ε	ε^*	ε^2}		
A''	1	1	1	1	1	−1	−1	−1	−1	−1	z	
E_1''	{1	ε	ε^2	ε^{2*}	ε^*	−1	$-\varepsilon$	$-\varepsilon^2$	$-\varepsilon^{2*}$	$-\varepsilon^*$}	(R_x, R_y)	(xz, yz)
	{1	ε^*	ε^{2*}	ε^2	ε	−1	$-\varepsilon^*$	$-\varepsilon^{2*}$	$-\varepsilon^2$	$-\varepsilon$}		
E_2''	{1	ε^2	ε^*	ε	ε^{2*}	−1	$-\varepsilon^2$	$-\varepsilon^*$	$-\varepsilon$	$-\varepsilon^{2*}$}		
	{1	ε^{2*}	ε	ε^*	ε^2	−1	$-\varepsilon^{2*}$	$-\varepsilon$	$-\varepsilon^*$	$-\varepsilon^2$}		

5. THE C_{nh} GROUPS *(continued)*

C_{6h}	E	C_6	C_3	C_2	C_3^2	C_6^5	i	S_3^5	S_6^5	σ_h	S_6	S_3		$\varepsilon = \exp(2\pi i/6)$
A_g	1	1	1	1	1	1	1	1	1	1	1	1	R_z	$x^2 + y^2, z^2$
B_g	1	-1	1	-1	1	-1	1	-1	1	-1	1	-1		
E_{1g}	1	ε	$-\varepsilon^*$	-1	$-\varepsilon$	ε^*	1	ε	$-\varepsilon^*$	-1	$-\varepsilon$	ε^*	(R_x, R_y)	(xz, yz)
	1	ε^*	$-\varepsilon$	-1	$-\varepsilon^*$	ε	1	ε^*	$-\varepsilon$	-1	$-\varepsilon^*$	ε		
E_{2g}	1	$-\varepsilon^*$	$-\varepsilon$	1	$-\varepsilon^*$	$-\varepsilon$	1	$-\varepsilon^*$	$-\varepsilon$	1	$-\varepsilon^*$	$-\varepsilon$		$(x^2 - y^2, xy)$
	1	$-\varepsilon$	$-\varepsilon^*$	1	$-\varepsilon$	$-\varepsilon^*$	1	$-\varepsilon$	$-\varepsilon^*$	1	$-\varepsilon$	$-\varepsilon^*$		
A_u	1	1	1	1	1	1	-1	-1	-1	-1	-1	-1	z	
B_u	1	-1	1	-1	1	-1	-1	1	-1	1	-1	1		
E_{1u}	1	ε	$-\varepsilon^*$	-1	$-\varepsilon$	ε^*	-1	$-\varepsilon$	ε^*	1	ε	$-\varepsilon^*$	(x, y)	
	1	ε^*	$-\varepsilon$	-1	$-\varepsilon^*$	ε	-1	$-\varepsilon^*$	ε	1	ε^*	$-\varepsilon$		
E_{2u}	1	$-\varepsilon^*$	$-\varepsilon$	1	$-\varepsilon^*$	$-\varepsilon$	-1	ε^*	ε	-1	ε^*	ε		
	1	$-\varepsilon$	$-\varepsilon^*$	1	$-\varepsilon$	$-\varepsilon^*$	-1	ε	ε^*	-1	ε	ε^*		

6. THE D_{nh} GROUPS

D_{2h}	E	$C_2(z)$	$C_2(y)$	$C_2(x)$	i	$\sigma(xy)$	$\sigma(xz)$	$\sigma(yz)$		
A_g	1	1	1	1	1	1	1	1		x^2, y^2, z^2
B_{1g}	1	1	-1	-1	1	1	-1	-1	R_z	xy
B_{2g}	1	-1	1	-1	1	-1	1	-1	R_y	xz
B_{3g}	1	-1	-1	1	1	-1	-1	1	R_x	yz
A_u	1	1	1	1	-1	-1	-1	-1		
B_{1u}	1	1	-1	-1	-1	-1	1	1	z	
B_{2u}	1	-1	1	-1	-1	1	-1	1	y	
B_{3u}	1	-1	-1	1	-1	1	1	-1	x	

6. THE D_{nh} GROUPS (continued)

D_{3h}	E	$2C_3$	$3C_2$	σ_h	$2S_3$	$3\sigma_v$		
A'_1	1	1	1	1	1	1		x^2+y^2, z^2
A'_2	1	1	−1	1	1	−1	R_z	
E'	2	−1	0	2	−1	0	(x, y)	(x^2-y^2, xy)
A''_1	1	1	1	−1	−1	−1		
A''_2	1	1	−1	−1	−1	1	z	
E''	2	−1	0	−2	1	0	(R_x, R_y)	(xz, yz)

D_{4h}	E	$2C_4$	C_2	$2C'_2$	$2C''_2$	i	$2S_4$	σ_h	$2\sigma_v$	$2\sigma_d$		
A_{1g}	1	1	1	1	1	1	1	1	1	1		x^2+y^2, z^2
A_{2g}	1	1	1	−1	−1	1	1	1	−1	−1	R_z	
B_{1g}	1	−1	1	1	−1	1	−1	1	1	−1		x^2-y^2
B_{2g}	1	−1	1	−1	1	1	−1	1	−1	1		xy
E_g	2	0	−2	0	0	2	0	−2	0	0	(R_x, R_y)	(xz, yz)
A_{1u}	1	1	1	1	1	−1	−1	−1	−1	−1		
A_{2u}	1	1	1	−1	−1	−1	−1	−1	1	1	z	
B_{1u}	1	−1	1	1	−1	−1	1	−1	−1	1		
B_{2u}	1	−1	1	−1	1	−1	1	−1	1	−1		
E_u	2	0	−2	0	0	−2	0	2	0	0	(x, y)	

6. THE D_{nh} GROUPS *(continued)*

D_{5h}	E	$2C_5$	$2C_5^2$	$5C_2$	σ_h	$2S_5$	$2S_5^3$	$5\sigma_v$		
A_1'	1	1	1	1	1	1	1	1		$x^2 + y^2, z^2$
A_2'	1	1	1	−1	1	1	1	−1	R_z	
E_1'	2	2 cos 72°	2 cos 144°	0	2	2 cos 72°	2 cos 144°	0	(x, y)	
E_2'	2	2 cos 144°	2 cos 72°	0	2	2 cos 144°	2 cos 72°	0		$(x^2 - y^2, xy)$
A_1''	1	1	1	1	−1	−1	−1	−1		
A_2''	1	1	1	−1	−1	−1	−1	1	z	
E_1''	2	2 cos 72°	2 cos 144°	0	−2	−2 cos 72°	−2 cos 144°	0	(R_x, R_y)	(xz, yz)
E_2''	2	2 cos 144°	2 cos 72°	0	−2	−2 cos 144°	−2 cos 72°	0		

D_{6h}	E	$2C_6$	$2C_3$	C_2	$3C_2'$	$3C_2''$	i	$2S_3$	$2S_6$	σ_h	$3\sigma_d$	$3\sigma_v$		
A_{1g}	1	1	1	1	1	1	1	1	1	1	1	1		$x^2 + y^2, z^2$
A_{2g}	1	1	1	1	−1	−1	1	1	1	1	−1	−1	R_z	
B_{1g}	1	−1	1	−1	1	−1	1	−1	1	−1	1	−1		
B_{2g}	1	−1	1	−1	−1	1	1	−1	1	−1	−1	1		
E_{1g}	2	1	−1	−2	0	0	2	1	−1	−2	0	0	(R_x, R_y)	(xz, yz)
E_{2g}	2	−1	−1	2	0	0	2	−1	−1	2	0	0		$(x^2 - y^2, xy)$
A_{1u}	1	1	1	1	1	1	−1	−1	−1	−1	−1	−1		
A_{2u}	1	1	1	1	−1	−1	−1	−1	−1	−1	1	1	z	
B_{1u}	1	−1	1	−1	1	−1	−1	1	−1	1	−1	1		
B_{2u}	1	−1	1	−1	−1	1	−1	1	−1	1	1	−1		
E_{1u}	2	1	−1	−2	0	0	−2	−1	1	2	0	0	(x, y)	
E_{2u}	2	−1	−1	2	0	0	−2	1	1	−2	0	0		

7. THE D_{nd} GROUPS

D_{2d}	E	$2S_4$	C_2	$2C_2'$	$2\sigma_d$		
A_1	1	1	1	1	1		x^2+y^2, z^2
A_2	1	1	1	−1	−1	R_z	
B_1	1	−1	1	1	−1		x^2-y^2
B_2	1	−1	1	−1	1	z	xy
E	2	0	−2	0	0	$(x, y); (R_x, R_y)$	(xz, yz)

D_{3d}	E	$2C_3$	$3C_2$	i	$2S_6$	$3\sigma_d$		
A_{1g}	1	1	1	1	1	1		x^2+y^2, z^2
A_{2g}	1	1	−1	1	1	−1	R_z	
E_g	2	−1	0	2	−1	0	(R_x, R_y)	$(x^2-y^2, xy), (xz, yz)$
A_{1u}	1	1	1	−1	−1	−1		
A_{2u}	1	1	−1	−1	−1	1	z	
E_u	2	−1	0	−2	1	0	(x, y)	

D_{4d}	E	$2S_8$	$2C_4$	$2S_8^3$	C_2	$4C_2'$	$4\sigma_d$		
A_1	1	1	1	1	1	1	1		x^2+y^2, z^2
A_2	1	1	1	1	1	−1	−1	R_z	
B_1	1	−1	1	−1	1	1	−1		
B_2	1	−1	1	−1	1	−1	1	z	
E_1	2	$\sqrt{2}$	0	$-\sqrt{2}$	−2	0	0	(x, y)	
E_2	2	0	−2	0	2	0	0		(x^2-y^2, xy)
E_3	2	$-\sqrt{2}$	0	$\sqrt{2}$	−2	0	0	(R_x, R_y)	(xz, yz)

7. THE D_{nd} GROUPS *(continued)*

D_{5d}	E	$2C_5$	$2C_5^2$	$5C_2$	i	$2S_{10}^3$	$2S_{10}$	$5\sigma_d$		
A_{1g}	1	1	1	1	1	1	1	1		x^2+y^2, z^2
A_{2g}	1	1	1	−1	1	1	1	−1	R_z	
E_{1g}	2	2 cos 72°	2 cos 144°	0	2	2 cos 72°	2 cos 144°	0	(R_x, R_y)	(xz, yz)
E_{2g}	2	2 cos 144°	2 cos 72°	0	2	2 cos 144°	2 cos 72°	0		(x^2-y^2, xy)
A_{1u}	1	1	1	1	−1	−1	−1	−1		
A_{2u}	1	1	1	−1	−1	−1	−1	1	z	
E_{1u}	2	2 cos 72°	2 cos 144°	0	−2	−2 cos 72°	−2 cos 144°	0	(x, y)	
E_{2u}	2	2 cos 144°	2 cos 72°	0	−2	−2 cos 144°	−2 cos 72°	0		

D_{6d}	E	$2S_{12}$	$2C_6$	$2S_4$	$2C_3$	$2S_{12}^5$	C_2	$6C_2'$	$6\sigma_d$		
A_1	1	1	1	1	1	1	1	1	1		x^2+y^2, z^2
A_2	1	1	1	1	1	1	1	−1	−1	R_z	
B_1	1	−1	1	−1	1	−1	1	1	−1		
B_2	1	−1	1	−1	1	−1	1	−1	1	z	
E_1	2	$\sqrt{3}$	1	0	−1	$-\sqrt{3}$	−2	0	0	(x, y)	
E_2	2	1	−1	−2	−1	1	2	0	0		(x^2-y^2, xy)
E_3	2	0	−2	0	2	0	−2	0	0		
E_4	2	−1	−1	2	−1	−1	2	0	0		
E_5	2	$-\sqrt{3}$	1	0	−1	$\sqrt{3}$	−2	0	0	(R_x, R_y)	(xz, yz)

8. THE S_n GROUPS

S_4	E	S_4	C_2	S_4^3		
A	1	1	1	1	R_z	x^2+y^2, z^2
B	1	-1	1	-1	z	x^2-y^2, xy
E	1	i	-1	$-i$	$(x, y); (R_x, R_y)$	(xz, yz)
	1	$-i$	-1	i		

S_6	E	C_3	C_3^2	i	S_6^5	S_6		$\varepsilon = \exp(2\pi i/3)$
A_g	1	1	1	1	1	1	R_z	x^2+y^2, z^2
E_g	1	ε	ε^*	1	ε	ε^*	(R_x, R_y)	$(x^2-y^2, xy); (xz, yz)$
	1	ε^*	ε	1	ε^*	ε		
A_u	1	1	1	-1	-1	-1	z	
E_u	1	ε	ε^*	-1	$-\varepsilon$	$-\varepsilon^*$	(x, y)	
	1	ε^*	ε	-1	$-\varepsilon^*$	$-\varepsilon$		

S_8	E	S_8	C_4	S_8^3	C_2	S_8^5	C_4^3	S_8^7		$\varepsilon = \exp(2\pi i/8)$
A	1	1	1	1	1	1	1	1	R_z	x^2+y^2, z^2
B	1	-1	1	-1	1	-1	1	-1	z	
E_1	1	ε	i	$-\varepsilon^*$	-1	$-\varepsilon$	$-i$	ε^*	$(x, y); (R_x, R_y)$	
	1	ε^*	$-i$	$-\varepsilon$	-1	$-\varepsilon^*$	i	ε		
E_2	1	i	-1	$-i$	1	i	-1	$-i$		(x^2-y^2, xy)
	1	$-i$	-1	i	1	$-i$	-1	i		
E_3	1	$-\varepsilon^*$	$-i$	ε	-1	ε^*	i	$-\varepsilon$		(xz, yz)
	1	$-\varepsilon$	i	ε^*	-1	ε	$-i$	$-\varepsilon^*$		

9. THE CUBIC GROUPS

T_d	E	$8C_3$	$3C_2$	$6S_4$	$6\sigma_d$		
A_1	1	1	1	1	1		$x^2 + y^2 + z^2$
A_2	1	1	1	−1	−1		
E	2	−1	2	0	0		$(2z^2 - x^2 - y^2, x^2 - y^2)$
T_1	3	0	−1	1	−1	(R_x, R_y, R_z)	
T_2	3	0	−1	−1	1	(x, y, z)	(xy, xz, yz)

O_h	E	$8C_3$	$6C_2$	$6C_4$	$3C_2 (= C_4^2)$	i	$6S_4$	$8S_6$	$3\sigma_h$	$6\sigma_d$		
A_{1g}	1	1	1	1	1	1	1	1	1	1		$x^2 + y^2 + z^2$
A_{2g}	1	1	−1	−1	1	1	−1	1	1	−1		
E_g	2	−1	0	0	2	2	0	−1	2	0		$(2z^2 - x^2 - y^2, x^2 - y^2)$
T_{1g}	3	0	−1	1	−1	3	1	0	−1	−1	(R_x, R_y, R_z)	
T_{2g}	3	0	1	−1	−1	3	−1	0	−1	1		(xz, yz, xy)
A_{1u}	1	1	1	1	1	−1	−1	−1	−1	−1		
A_{2u}	1	1	−1	−1	1	−1	1	−1	−1	1		
E_u	2	−1	0	0	2	−2	0	1	−2	0		
T_{1u}	3	0	−1	1	−1	−3	−1	0	1	1	(x, y, z)	
T_{2u}	3	0	1	−1	−1	−3	1	0	1	−1		

10. THE GROUPS $C_{\infty v}$ AND $D_{\infty h}$

$C_{\infty v}$	E	$2C_\infty^\Phi$	$\cdots$	$\infty\sigma_v$		
$A_1 \equiv \Sigma^+$	1	1	$\cdots$	1	z	$x^2 + y^2, z^2$
$A_2 \equiv \Sigma^-$	1	1	$\cdots$	-1	R_z	
$E_1 \equiv \Pi$	2	$2 \cos \Phi$	$\cdots$	0	$(x, y); (R_x, R_y)$	(xz, yz)
$E_2 \equiv \Delta$	2	$2 \cos 2\Phi$	$\cdots$	0		$(x^2 - y^2, xy)$
$E_3 \equiv \Phi$	2	$2 \cos 3\Phi$	$\cdots$	0		
$\cdots$	$\cdots$	$\cdots$	$\cdots$	$\cdots$		

$D_{\infty h}$	E	$2C_\infty^\Phi$	$\cdots$	$\infty\sigma_v$	i	$2S_\infty^\Phi$	$\cdots$	∞C_2		
Σ_g^+	1	1	$\cdots$	1	1	1	$\cdots$	1		$x^2 + y^2, z^2$
Σ_g^-	1	1	$\cdots$	-1	1	1	$\cdots$	-1	R_z	
Π_g	2	$2 \cos \Phi$	$\cdots$	0	2	$-2 \cos \Phi$	$\cdots$	0	(R_x, R_y)	(xz, yz)
Δ_g	2	$2 \cos 2\Phi$	$\cdots$	0	2	$2 \cos 2\Phi$	$\cdots$	0		$(x^2 - y^2, xy)$
$\cdots$	$\cdots$	$\cdots$	$\cdots$	$\cdots$	$\cdots$	$\cdots$	$\cdots$	$\cdots$		
Σ_u^+	1	1	$\cdots$	1	-1	-1	$\cdots$	-1	z	
Σ_u^-	1	1	$\cdots$	-1	-1	-1	$\cdots$	1		
Π_u	2	$2 \cos \Phi$	$\cdots$	0	-2	$2 \cos \Phi$	$\cdots$	0	(x, y)	
Δ_u	2	$2 \cos 2\Phi$	$\cdots$	0	-2	$-2 \cos 2\Phi$	$\cdots$	0		
$\cdots$	$\cdots$	$\cdots$	$\cdots$	$\cdots$	$\cdots$	$\cdots$	$\cdots$	$\cdots$		

11. THE ICOSAHEDRAL GROUP

I_h	E	$12C_5$	$12C_5^2$	$20C_3$	$15C_2$	i	$12S_{10}$	$12S_{10}^3$	$20S_6$	15σ		
A_g	1	1	1	1	1	1	1	1	1	1		$x^2+y^2+z^2$
T_{1g}	3	$\frac{1}{2}(1+\sqrt{5})$	$\frac{1}{2}(1-\sqrt{5})$	0	-1	3	$\frac{1}{2}(1-\sqrt{5})$	$\frac{1}{2}(1+\sqrt{5})$	0	-1	(R_x, R_y, R_z)	
T_{2g}	3	$\frac{1}{2}(1-\sqrt{5})$	$\frac{1}{2}(1+\sqrt{5})$	0	-1	3	$\frac{1}{2}(1+\sqrt{5})$	$\frac{1}{2}(1-\sqrt{5})$	0	-1		
G_g	4	-1	-1	1	0	4	-1	-1	1	0		
H_g	5	0	0	-1	1	5	0	0	-1	1		$(2z^2-x^2-y^2,$ $x^2-y^2,$ $xy, yz, xz)$
A_u	1	1	1	1	1	-1	-1	-1	-1	-1		
T_{1u}	3	$\frac{1}{2}(1+\sqrt{5})$	$\frac{1}{2}(1-\sqrt{5})$	0	-1	-3	$-\frac{1}{2}(1-\sqrt{5})$	$-\frac{1}{2}(1+\sqrt{5})$	0	1	(x, y, z)	
T_{2u}	3	$\frac{1}{2}(1-\sqrt{5})$	$\frac{1}{2}(1+\sqrt{5})$	0	-1	-3	$-\frac{1}{2}(1+\sqrt{5})$	$-\frac{1}{2}(1-\sqrt{5})$	0	1		
G_u	4	-1	-1	1	0	-4	1	1	-1	0		
H_u	5	0	0	-1	1	-5	0	0	1	-1		

Appendix II

SELECTED CORRELATION TABLES

D_{4h}	D_4	$C_2' \rightarrow C_2'$ D_{2d}	$C_2'' \rightarrow C_2'$ D_{2d}	C_{4v}	C_{4h}	C_2' D_{2h}	C_2'' D_{2h}	C_4	S_4
A_{1g}	A_1	A_1	A_1	A_1	A_g	A_g	A_g	A	A
A_{2g}	A_2	A_2	A_2	A_2	A_g	B_{1g}	B_{1g}	A	A
B_{1g}	B_1	B_1	B_2	B_1	B_g	A_g	B_{1g}	B	B
B_{2g}	B_2	B_2	B_1	B_2	B_g	B_{1g}	A_g	B	B
E_g	E	E	E	E	E_g	$B_{2g} + B_{3g}$	$B_{2g} + B_{3g}$	E	E
A_{1u}	A_1	B_1	B_1	A_2	A_u	A_u	A_u	A	B
A_{2u}	A_2	B_2	B_2	A_1	A_u	B_{1u}	B_{1u}	A	B
B_{1u}	B_1	A_1	A_2	B_2	B_u	A_u	B_{1u}	B	A
B_{2u}	B_2	A_2	A_1	B_1	B_u	B_{1u}	A_u	B	A
E_u	E	E	E	E	E_u	$B_{2u} + B_{3u}$	$B_{2u} + B_{3u}$	E	E

D_{4h} (cont.)	C'_2 D_2	C''_2 D_2	C_2, σ_v C_{2v}	C_2, σ_d C_{2v}	C'_2 C_{2v}	C''_2 C_{2v}
A_{1g}	A	A	A_1	A_1	A_1	A_1
A_{2g}	B_1	B_1	A_2	A_2	B_1	B_1
B_{1g}	A	B_1	A_1	A_2	A_1	B_1
B_{2g}	B_1	A	A_2	A_1	B_1	A_1
E_g	$B_2 + B_3$	$B_2 + B_3$	$B_1 + B_2$	$B_1 + B_2$	$A_2 + B_2$	$A_2 + B_2$
A_{1u}	A	A	A_2	A_2	A_2	A_2
A_{2u}	B_1	B_1	A_1	A_1	B_2	B_2
B_{1u}	A	B_1	A_2	A_1	A_2	B_2
B_{2u}	B_1	A	A_1	A_2	B_2	A_2
E_u	$B_2 + B_3$	$B_2 + B_3$	$B_1 + B_2$	$B_1 + B_2$	$A_1 + B_1$	$A_1 + B_1$

D_{4h} (cont.)	C_2 C_{2h}	C'_2 C_{2h}	C''_2 C_{2h}	C_2 C_2	C'_2 C_2	C''_2 C_2	σ_h C_s	σ_v C_s	σ_d C_s	C_i
A_{1g}	A_g	A_g	A_g	A	A	A	A'	A'	A'	A_g
A_{2g}	A_g	B_g	B_g	A	B	B	A'	A''	A''	A_g
B_{1g}	A_g	A_g	B_g	A	A	B	A'	A'	A''	A_g
B_{2g}	A_g	B_g	A_g	A	B	A	A'	A''	A'	A_g
E_g	$2B_g$	$A_g + B_g$	$A_g + B_g$	$2B$	$A + B$	$A + B$	$2A''$	$A' + A''$	$A' + A''$	$2A_g$
A_{1u}	A_u	A_u	A_u	A	A	A	A''	A''	A''	A_u
A_{2u}	A_u	B_u	B_u	A	B	B	A''	A'	A'	A_u
B_{1u}	A_u	A_u	B_u	A	A	B	A''	A''	A'	A_u
B_{2u}	A_u	B_u	A_u	A	3	A	A''	A'	A''	A_u
E_u	$2B_u$	$A_u + B_u$	$A_u + B_u$	$2B$	$A + B$	$A + B$	$2A'$	$A' + A''$	$A' + A''$	$2A_u$

T_d	T	D_{2d}	C_{3v}	S_4	D_2	C_{2v}	C_3	C_2	C_s
A_1	A	A_1	A_1	A	A	A_1	A	A	A'
A_2	A	B_1	A_2	B	A	A_2	A	A	A''
E	E	$A_1 + B_1$	E	$A + B$	$2A$	$A_1 + A_2$	E	$2A$	$A' + A''$
T_1	T	$A_2 + E$	$A_2 + E$	$A + E$	$B_1 + B_2 + B_3$	$A_2 + B_1 + B_2$	$A + E$	$A + 2B$	$A' + 2A''$
T_2	T	$B_2 + E$	$A_1 + E$	$B + E$	$B_1 + B_2 + B_3$	$A_1 + B_1 + B_2$	$A + E$	$A + 2B$	$2A' + A''$

O	T	D_4	D_3	C_4	$3C_2$ D_2	$C_2, 2C'_2$ D_2	C_3	C_2	C_2
A_1	A	A_1	A_1	A	A	A	A	A	A
A_2	A	B_1	A_2	B	A	B_1	A	A	B
E	E	$A_1 + B_1$	E	$A + B$	$2A$	$A + B_1$	E	$2A$	$A + B$
T_1	T	$A_2 + E$	$A_2 + E$	$A + E$	$B_1 + B_2 + B_3$	$B_1 + B_2 + B_3$	$A + E$	$A + 2B$	$A + 2B$
T_2	T	$B_2 + E$	$A_1 + E$	$B + E$	$B_1 + B_2 + B_3$	$A + B_2 + B_3$	$A + E$	$A + 2B$	$2A + B$

O_h	O	T_d	T_h	D_{4h}	D_{3d}
A_{1g}	A_1	A_1	A_g	A_{1g}	A_{1g}
A_{2g}	A_2	A_2	A_g	B_{1g}	A_{2g}
E_g	E	E	E_g	$A_{1g} + B_{1g}$	E_g
T_{1g}	T_1	T_1	T_g	$A_{2g} + E_g$	$A_{2g} + E_g$
T_{2g}	T_2	T_2	T_g	$B_{2g} + E_g$	$A_{1g} + E_g$
A_{1u}	A_1	A_2	A_u	A_{1u}	A_{1u}
A_{2u}	A_2	A_1	A_u	B_{1u}	A_{2u}
E_u	E	E	E_u	$A_{1u} + B_{1u}$	E_u
T_{1u}	T_1	T_2	T_u	$A_{2u} + E_u$	$A_{2u} + E_u$
T_{2u}	T_2	T_1	T_u	$B_{2u} + E_u$	$A_{1u} + E_u$

Appendix III

ATOMIC UNITS

Quantity	*Size of atomic unit in c.g.s. units*	*Value in atomic units*
Mass	$m_e = 9.1091 \times 10^{-28}$ gm	Mass of electron = 1 Mass of proton = 1836.1
Length	$a_0 = \hbar^2/m_e e^2$ $= 0.529167 \times 10^{-8}$ cm.	Radius of innermost Bohr orbit = 1
Time	$\tau_0 = a_0\hbar/e^2$ $= 2.4189 \times 10^{-17}$ sec.	Time for one circuit of innermost Bohr orbit = 2π
Velocity	$e^2/\hbar = 2.1877 \times 10^8$ cm./sec.	Velocity of electron in innermost Bohr orbit = 1 Velocity of light = 137.037
Action	$h = 6.6256 \times 10^{-27}$ erg-sec.	Planck's constant = 1
Angular momentum	$\hbar = h/2\pi$ $= 1.05450 \times 10^{-27}$ erg-sec.	
Energy	$e^2/a_0 = 4.3592 \times 10^{-11}$ erg	Ionization energy of hydrogen = 1/2
Electric charge	$e = 4.8029 \times 10^{-10}$ e.s.u. $= 1.6021 \times 10^{-20}$ e.m.u.	charge on electron = −1
Magnetic moment	$e\hbar/2m_e c = 0.92732 \times 10^{-20}$ erg/gauss = 1 Bohr magneton	Magnetic moment due to orbital motion in innermost Bohr orbit = 1

INDEX

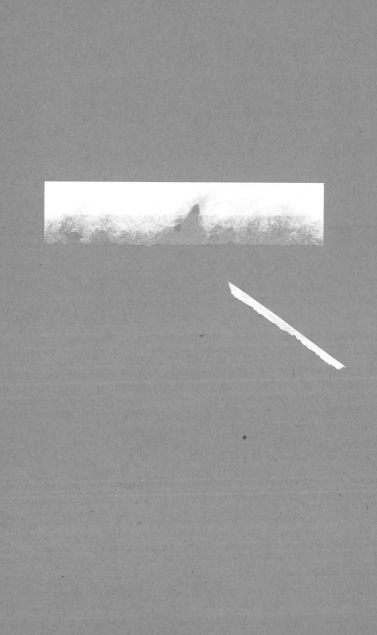